BASIC CHEMISTRY CALCULATIONS:

A Book for Chemistry

and

Chemical Engineering Students

By

Kingsley Augustine

TABLE OF CONTENT

CHAPTER 1 MOLE FRACTION AND MASS FRACTION 3

CHAPTER 2 AVERAGE MOLECULAR MASS 13

CHAPTER 3 CALCULATIONS INVOLVING COMBUSTION 20

CHAPTER 4 CALCULATIONS INVOLVING LIMITING REACTANTS 39

CHAPTER 5 CALCULATIONS INVOLVING THE FORMULA OF COMPOUNDS 53

CHAPTER 6 EQUILIBRIUM REACTION CALCULATIONS 66

CHAPTER 1
MOLE FRACTION AND MASS FRACTION

The number of moles of a substance is the ratio of the mass of the substance to the molar mass of the substance. It is expressed as:

$$\text{Number of moles} = \frac{\text{mass of substance}}{\text{molar mass of substance}}$$

Mole fraction is the ratio of the number of moles of any component in a mixture to the total number of moles of all the components in the mixture. It is given by:

$$\text{Mole ratio, } y = \frac{\text{number of moles of any component}}{\text{total number of moles of all components}}$$

Mass fraction is the ratio of the mass of any component in a mixture to the total mass of all the components in the mixture. It is given by:

$$\text{Mass fraction, } x = \frac{\text{mass of any component}}{\text{total mass of all the components}}$$

Note that the sum of the mole fractions of all the components in a mixture is equal to one.

Similarly, the sum of the mass fractions of all the components in a mixture is equal to one

Thus, if y is the mole fraction and x is a mass fraction, then symbolically, we have that:

$$\sum y = 1 \quad \text{and} \quad \sum x = 1$$

where the symbol, $\sum$, means summation.

Examples

1. A mixture contains 4.2 kg of X, 1.8 kg of Y and 3.5 kg of Z. The molecular mass of X, Y and Z are 16kg/kmol, 44kg/kmol and 28kg/kmol respectively. Calculate:

(a). The number of moles of each component

(b). The mole fraction of each component

(c). The mass fraction of each component

Solutions

(a). Number of kmols = $\dfrac{\text{mass of component}}{\text{molar mass of component}}$

Therefore, number of kmols of X = $\dfrac{4.2}{16}$

$\qquad\qquad\qquad\qquad$ = 0.2625kmols

Number of kmols of Y = $\dfrac{1.8}{44}$

$\qquad\qquad\qquad\qquad$ = 0.0409kmols

Number of kmols of Z = $\dfrac{3.5}{28}$

$\qquad\qquad\qquad\qquad$ = 0.1250kmols

(b). Total kmols of all components = 0.2625 + 0.0409 + 0.125

$\qquad\qquad\qquad\qquad$ = 0.4284

Mole fraction = $\dfrac{\text{Number of moles of a component}}{\text{total number on moles of all components}}$

Therefore, mole fraction of X = $\dfrac{0.2625}{0.4284}$

$\qquad\qquad\qquad\qquad$ = 0.613

Mole fraction of Y = $\dfrac{0.0409}{0.4284}$

$\qquad\qquad\qquad\qquad$ = 0.095

mole fraction of Z = $\dfrac{0.125}{0.4284}$

$\qquad\qquad\qquad\qquad$ = 0.292

(c). The total mass of all components = 4.2 + 1.8 + 3.5

$\qquad\qquad\qquad\qquad$ = 9.5kg

Therefore mass fraction of X = $\dfrac{4.2}{9.5}$

$\qquad\qquad\qquad\qquad$ = 0.442

Mass fraction of Y = $\dfrac{1.8}{9.5}$

= 0.189

Mass fraction of X = $\dfrac{3.5}{9.5}$

= 0.368

2. A Liquefied petroleum gas (LPG) was analyzed to contain the following composition by volume: 52.4% CH_4, 30.9% C_4H_{10} and 16.7% C_3H_8. Calculate:

(a). the mole fraction of each component

(b). the mass of each component

(c). the mass fraction of each component

Solutions

(a). The composition by volume of the component is also regarded as the number of moles of the components.

Therefore, the LPG contains 52.4 moles of CH_4, 30.9 moles of C_4H_{10} and 16.7 moles of C_3H_8

Total number of moles = 100 moles (i.e. 30.9 + 52.4 + 16.7 = 100)

Therefore mole fraction of CH_4 = $\dfrac{52.4}{100}$

= 0.524

Mole fraction of C_4H_{10} = $\dfrac{30.9}{100}$

= 0.309

Mole fraction of C_3H_8 = $\dfrac{16.7}{100}$

= 0.167

(b). Recall that: Number of moles = $\dfrac{Mass}{Molar\ mass}$

Therefore, mass = molar mass × number of moles   (when we cross multiply)

Therefore mass of CH_4 = (12 + 4) x 52.4   (12 + 4 = 16 = molar mass of CH_4)

= 16 x 52.4

= 838.4g

Mass of C_4H_{10} = [(12 x 4) + (1 x 10)] x 30.9

= 58 x 30.9   (Molar mass of C_4H_{10} = 58)

= 1792.2g

Mass of C_3H_8 = [(12 x 3) + (1 x 8)] x 16.7

= 44 x 16.7

= 734.8g

(c). Total mass of the hydrocarbon = 838.4 + 1792.2 + 734.8

= 3365.4g

Therefore, mass fraction of $CH_4 = \dfrac{838.4}{3365.4}$

= 0.249

Mass fraction of $C_4H_{10} = \dfrac{1792.2}{3365.4}$

= 0.533

Mass fraction of $C_3H_8 = \dfrac{734.8}{3365.4}$

= 0.218

3. A hydrocarbon was burnt in excess air to produce a mixture of gases with the following composition: 1.24 moles of CO_2, 0.28mole of CO, 0.91 mole of O_2 and 5.66 of N_2. Determine the composition by mass of each component.

Solution

Mass of component = number of moles x molar mass

Therefore mass of CO_2 = 1.24 x [12 + (6 x 2)]

$$= 1.24 \times 44$$
$$= 54.56g$$

Mass of CO $= 0.28 \times (12 + 16)$
$$= 0.28 \times 28$$
$$= 7.84g$$

Mass of $O_2 = 0.91 \times (16 \times 2)$
$$= 0.91 \times 32$$
$$= 29.12g$$

Mass of $N_2 = 5.66 \times (14 \times 2)$
$$= 5.66 \times 28$$
$$= 158.48g$$

Note that the mass fraction of each component can be calculated as follows:

Total mass of all components $= 54.56 + 7.84 + 29.12 + 158.48$
$$= 250g$$

Therefore mass fraction of $CO_2 = \dfrac{54.56}{250}$
$$= 0.218$$

Mass fraction of CO $= \dfrac{7.84}{250}$
$$= 0.031$$

Mass fraction of $O_2 = \dfrac{29.12}{250}$
$$= 0.116$$

Mass fraction of $N_2 = \dfrac{158.48}{250}$
$$= 0.634$$

4. On analysis, a gaseous mixture of CO_2 and CO was found to weigh 138g. If the total number of moles of the mixture is 4.4 moles, calculate:

(a). The mass of each of CO_2 and CO in the mixture

(b). The mole fraction of each of the components

Solution

(a). Let the mass of CO_2 in the mixture be m. Therefore the mass of CO will be 138 - m

The molecular mass of CO_2 = 12 + (16 x 2)

$$= 12 + 32$$

$$= 44 \text{g/mol} \quad \text{(It can also be expressed as 44kg/kmol)}$$

The molecular mass of CO = 12 + 16 = 28g/mol

Therefore, the number of moles of $CO_2 = \dfrac{m}{44}$ (since mass of CO_2 = m)

The number of moles of CO = $\dfrac{138 - m}{28}$

But the total number of moles is 4.4moles. This means that:

$$\dfrac{m}{44} + \dfrac{138 - m}{28} = 4.4$$

In order to solve this equation, multiply each term by 308, (i.e. the L.C.M of 44 and 28). This gives:

$$308\left(\dfrac{m}{44}\right) + 308\left(\dfrac{138 - m}{28}\right) = 308 \times 4.4$$

$7m + 11(138 - m) = 1355.2$ (Note that $\dfrac{308}{44}$ = 7, and $\dfrac{308}{28}$ = 11)

$7m + 1518 - 11m = 1355.2$

$1518 - 1355.2 = 11m - 7m$

$162.8 = 4m$

Therefore, $m = \dfrac{162.8}{4}$

$m = 40.7\text{g}$

Hence, the mass of CO_2 in the mixture is 40.7g

And the mass of CO in the mixture is 138 - m = 138 - 40.7

$$= 97.3g$$

(b). Number of moles of CO_2 = 40.7/44 = 0.925 moles

Number of moles of CO = $\dfrac{97.3}{28}$

$$= 3.475 \text{ moles}$$

The total number of moles = 4.4 moles

Therefore mole fraction of CO_2, y_{CO_2} = $\dfrac{0.925}{4.4}$

$$= 0.210$$

Mole fraction of CO, y_{CO} = $\dfrac{3.475}{4.4}$

$$= 0.790$$

5. 45g of a gaseous mixture has the following composition: A = 0.75 moles and a mass fraction of 0.22, B = 0.32 moles and a mass fraction of 0.48, C = 0.94 moles and a mass fraction of 0.3. Calculate the molecular mass of each of A, B and C.

Solutions

Mass fraction = $\dfrac{\text{Mass of component}}{\text{Total mass}}$

Therefore, for A, $0.22 = \dfrac{\text{mass of A}}{45}$

Hence,   mass of A = 0.22 x 45   (when we cross multiply the above equation)

$$= 9.9g$$

Similarly,   mass of B = 0.48 x 45

$$= 21.6g$$

And,   mass of C = 0.3 x 45

= 13.5g

Recall that: Number of moles = $\dfrac{\text{Mass of substance}}{\text{Molar mass of substance}}$

Therefore, for A, $0.75 = \dfrac{9.9}{\text{molecular mass}}$

Hence, molecular mass of A = $\dfrac{9.9}{0.75}$

= 13.2g/mol

Similarly, molecular mass of B = $\dfrac{\text{mass of B}}{\text{number of moles of B}}$

= $\dfrac{21.6}{0.32}$

= 67.5g/mol

And, molecular mass of C = $\dfrac{13.5}{0.94}$

= 14.4g/mol

6. A glucose solution contains 32wt % of glucose ($C_6H_{12}O_6$). Calculate the mole fraction of glucose in the solution.

Solution

From the question, we can deduce that for a 100g (i.e. from 100%) of solution, the mass of glucose is 32g, while the mass of water is 100 - 32 = 68g.

Therefore, number of mole of glucose = $\dfrac{\text{mass of glucose}}{\text{molar mass of glucose}}$

= $\dfrac{32}{[(12 \times 6) + (1 \times 12) + (16 \times 6)]}$

= $\dfrac{32}{72 + 12 + 96)}$

= $\dfrac{32}{180}$

= 0.1778

Similarly, number of moles of water = $\frac{68}{18}$ (Molecular mass of water = 18)

$$= 3.778$$

Therefore, mole fraction of glucose = $\frac{0.1778}{0.1778 + 3.778}$

$$= \frac{0.1778}{3.9558}$$

$$= 0.0449$$

Therefore mole fraction of glucose in the solution is 0.0449.

EXERCISE

1. A mixture contains 5.3 kg of X, 2.5 kg of Y and 4.1 kg of Z. The molecular mass of X, Y and Z are 23kg/kmol, 31kg/kmol and 39kg/kmol respectively. Calculate:

(a). The number of moles of each component

(b). The mole fraction of each component

(c). The mass fraction of each component

2. A fuel was analyzed to contain the following composition by volume: 42.2% CH_4, 40.9% C_4H_{10} and 16.9% C_3H_8. Calculate:

(a). the mole fraction of each component

(b). the mass of each component

(c). the mass fraction of each component

3. A hydrocarbon was burnt in excess air to produce a mixture of gases with the following composition: 2.4 moles of CO_2, 1.5moles of CO, 3.2 moles of O_2 and 6.8 moles of N_2. Determine the composition by mass of each component.

4. On analysis, a gaseous mixture of NO_2 and NO was found to weigh 92g. If the total number of moles of the mixture is 3.6 moles, calculate:

(a). The mass of each of NO_2 and NO in the mixture

(b). The mole fraction of each of the component

5. 120g of a gaseous mixture has the following composition: X =1.55 moles and a mass fraction of 0.12, Y = 0.81 moles and a mass fraction of 0.55, Z = 2.32 moles and a mass fraction of 0.33. Calculate the molecular mass of each of X, Y and Z.

6. A solution of tetraoxosulphate (VI) acid contains 68wt % of the acid. Calculate the mole fraction of the acid in the solution.
(H = 1, S= 32, O = 16)

CHAPTER 2
AVERAGE MOLECULAR MASS

The average molecular mass of a mixture is the ratio of the mass of the mixture to the number of moles of all components in the mixture. It is given in terms of the mole fraction and molecular mass of each component as follows:

$M_{av} = y_1M_1 + y_2M_2 + y_3M_3 + \ldots y_tM_t$

where M_{av} = average molecular mass of mixture, y = mole fraction of component and M = molecular mass of component and there are t components in the mixture.

It can be represented symbolically as:

$M_{av} = \Sigma yM$

The average molecular mass of a mixture can also be given in terms of the mass fraction as follows:

$$\frac{1}{M_{av}} = \frac{x_1}{M_1} + \frac{x_2}{M_2} + \frac{x_3}{M_3} + \ldots \frac{x_t}{M_t}$$

where x = mass fraction of component.

Symbolically it can be expressed as:

$$\frac{1}{M_{av}} = \Sigma \frac{x}{M}$$

Examples

1. A mixture of gases is composed of 0.25 moles of N_2, 1.32 moles of CO and 0.71 moles of Ne. Calculate the average molecular mass of the mixture.

Solution

Total number of moles of the components = 0.25 + 1.32 + 0.71

$\qquad\qquad\qquad\qquad\qquad\qquad\qquad$ = 2.28

Therefore, mole fraction of $N_2 = \dfrac{0.25}{2.28}$

$\qquad\qquad\qquad\qquad\qquad\qquad$ = 0.110

Mole fraction of CO = $\dfrac{1.32}{2.28}$

$= 0.579$

Mole fraction of Ne = $\dfrac{0.71}{2.28}$

$= 0.311$

Therefore, $M_{av} = \sum yM$

$= (yM)_{N_2} + (yM)_{CO} + (yM)_{Ne}$

Note that the molecular mass of $N_2 = 14 \times 2 = 28$

The molecular mass of CO $= 12 + 16 = 28$

The atomic mass of Ne $= 20$

Therefore, $M_{av} = (yM)_{N_2} + (yM)_{CO} + (yM)_{Ne}$

$= (0.110 \times 28) + (0.579 \times 28) + (0.311 \times 20)$

$= 3.08 + 16.212 + 6.22$

$= 25.512$

Therefore, the average molecular mass of the mixture is 25.5g/mol

2. The components of a mixture are 1.5g of H_2, 0.95g of CO_2 and 0.22g of N_2. Determine the average molecular mass of the mixture.

Solution

Total mass of the components $= 1.5 + 0.95 + 0.22$

$= 2.67g$

Therefore, mass fraction of $H_2 = \dfrac{1.5}{2.67}$

$= 0.56$

Mass fraction of $CO_2 = \dfrac{0.95}{2.67}$

$$= 0.36$$

Mass fraction of $N_2 = \dfrac{0.22}{2.67}$

$$= 0.08$$

The molecular mass of $H_2 = 1 \times 2 = 2$

The molecular mass of $CO_2 = 12 + (16 \times 2)$

$$= 12 + 32 = 44$$

The molecular mass of $N_2 = 14 \times 2 = 28$

Therefore, $\dfrac{1}{M_{av}} = \sum \dfrac{x}{M}$

$$= \left(\dfrac{x}{M}\right)_{H_2} + \left(\dfrac{x}{M}\right)_{CO_2} + \left(\dfrac{x}{M}\right)_{N_2}$$

$$= \dfrac{0.56}{2} + \dfrac{0.36}{44} + \dfrac{0.08}{28}$$

$$= 0.28 + 0.0082 + 0.0029$$

$$\dfrac{1}{M_{av}} = 0.2911$$

Therefore, $M_{av} = \dfrac{1}{0.2911}$

$$M_{av} = 3.44$$

Therefore, the average molecular mass of the mixture is 3.44g/mol

3. A gaseous mixture contains O_2 and an unknown gas. If the average molecular mass of the mixture is 23g/mol and the mole fraction of O_2 in the mixture is 0.45, calculate the molecular mass of the unknown gas.

Solution

Recall that sum of mole fraction of all components of a mixture is = 1

Therefore, $y_{O_2} + y_A = 1$ (where A is the unknown gas)

$$0.45 + y_A = 1$$

$$y_A = 1 - 0.45$$

$$y_A = 0.55$$

Molecular mass of $O_2 = 16 \times 2 = 32$

Therefore, $M_{av} = \Sigma yM$

$$M_{av} = (yM)_{O_2} + (yM)_A$$

$$23 = (0.45 \times 32) + (0.55 M_A) \quad (M_A \text{ is the molecular mass of A})$$

$$23 = 14.4 + 0.55 M_A$$

$$23 - 14.4 = 0.55 M_A$$

$$8.6 = 0.55 M_A$$

Therefore, $M_A = \dfrac{8.6}{0.55}$

$$= 15.6$$

Therefore, the molecular mass of the unknown gas is 15.6g/mol

4. A gas is composed of N_2 and Ne. If the average molecular mass of the gas is 21.5g/mol, calculate the mass fraction of N_2 and Ne in the mixture.

Solution

Molecular mass of $N_2 = 14 \times 2 = 28$

Molecular mass of Ne = 20

Let the mass fraction of N_2 be x.

Therefore the mass fraction of Ne = 1 - x   (since sum of mass fractions is 1)

In terms of mass fraction:

$$\dfrac{1}{M_{av}} = \Sigma \dfrac{x}{M}$$

$$\frac{1}{M_{av}} = \left(\frac{x}{M}\right)_{N_2} + \left(\frac{x}{M}\right)_{Ne}$$

$$\frac{1}{21.5} = \frac{x}{28} + \frac{1-x}{20}$$

$$0.0465 = \frac{x}{28} + \frac{1-x}{20}$$

In order to clear the fractions, multiply each term by 140 (i.e. the LCM of 28 and 20)

$$140 \times 0.0465 = 140\left(\frac{x}{28}\right) + 140\left(\frac{1-x}{20}\right)$$

$$6.51 = 5x + 7(1 - x) \quad \text{(Note that } \frac{140}{28} = 5, \text{ and } \frac{140}{20} = 7\text{)}$$

$$6.51 = 5x + 7 - 7x$$

$$7x - 5x = 7 - 6.51$$

$$2x = 0.49$$

Therefore, $x = \frac{0.49}{2}$

$$x = 0.245$$

Therefore the mass fraction of N_2 = 0.245, while the mass fraction of Ne = 1 - x = 1 - 0.245 = 0.755.

5. The Composition of air by mass is 23.2% of O_2 and 76.8% of N_2, while the composition of air by volume is 21% of O_2 and 79% of N_2. Calculate the average molecular mass of air.

Solution

Method 1: Use of mass fraction.

Total mass percent of the components = 23.2 + 76.8 = 100

Therefore, mass fraction of $O_2 = \frac{23.2}{100}$

$$= 0.232$$

Mass fraction of $N_2 = \frac{76.8}{100}$

$$= 0.768$$

Therefore, $\dfrac{1}{M_{av}} = \left(\dfrac{x}{M}\right)_{O_2} + \left(\dfrac{x}{M}\right)_{N_2}$

$\qquad = \dfrac{0.232}{32} + \dfrac{0.768}{28}$ (Molecular mass of O_2 = 32, while that of N_2 = 28)

$\qquad = 0.00725 + 0.02743$

$\dfrac{1}{M_{av}} = 0.03468$

Therefore, $M_{av} = \dfrac{1}{0.03468}$

$M_{av} = 28.8 \text{g/mol}$

Method 2: Use of mole fraction.

Total mole (volume) percent of mixture components = 21 + 79 = 100

Therefore, mole fraction of $O_2 = \dfrac{21}{100}$

$\qquad = 0.21$

Mole fraction of $N_2 = \dfrac{79}{100}$

$\qquad = 0.79$

Therefore, $M_{av} = (yM)_{O_2} + (yM)_{N_2}$

$\qquad = (0.21 \times 32) + (0.79 \times 28)$

$\qquad = 6.72 + 22.12$

$M_{av} = 28.8$

Therefore the average molecular mass of air is 28.8g/mol.

EXERCISE

1. A mixture of gases is composed of 0.58 moles of CO_2, 1.95 moles of NO and 3.25 moles of O_2. Calculate the average molecular mass of the mixture.

2. The components of a mixture are 10g of H_2, 25g of NO_2 and 7.8g of N_2. Determine the average molecular mass of the mixture.

3. A gaseous mixture contains N_2 and an unknown gas. If the average molecular mass of the mixture is 52g/mol and the mole fraction of N_2 in the mixture is 0.85, calculate the molecular mass of the unknown gas.

4. A gas is composed of Cl_2 and Ar. If the average molecular mass of the gas is 71.6g/mol, calculate the mass fraction of Cl_2 and Ar in the mixture.

5. The Composition of air by mass is 23% of O_2 and 77% of N_2, while the composition of air by volume is 20.8% of O_2 and 79.2% of N_2. Calculate the average molecular mass of air.

6. A gaseous mixture contains O_2 and an unknown gas. If the average molecular mass of the mixture is 25.6g/mol and the mole fraction of O_2 in the mixture is 0.62, calculate the molecular mass of the unknown gas.

7. A gas is composed of SO_2 and He. If the average molecular mass of the gas is 91g/mol, calculate the mass fraction of SO_2 and He in the mixture.

8. The Composition of an inert gas combination by volume is 27.2% of He, 35.6% of Ne and 37.2% of Ar. Calculate the average molecular mass of this combination.

CHAPTER 3
CALCULATIONS INVOLVING COMBUSTION

Some of the terms associated with combustion are:

(I). Stack or flue gas: This involves all the gases that result from a combustion process. Flue or stack gas contains water vapour, hence it is a wet gas.

(II). Orsat analysis: This is a dry gas analysis because it contains all the gases excluding water vapour from a combustion process.

In combustion process calculations, excess air or oxygen and theoretical air or oxygen are usually calculated.

Examples

1. 20kg of methane is combusted with 380kg of air to give 36kg of carbon (IV) oxide and 14kg of carbon (II) oxide. Determine the percent excess air used.

(Air contains 23.2% by mass of oxygen, and the molecular mass of air is 29kg/kmol)

Solution

Since excess air was supplied, it means that the reaction is a complete combustion. The first step is to calculate the amount of oxygen needed for the combustion. This is obtained from the complete combustion reaction.

The balanced equation for the complete combustion reaction is:

$$CH_4 + 2O_2 \longrightarrow CO_2 + 2H_2O$$
$$16kg + 64kg \longrightarrow 44kg + 36kg$$

Note that the molecular mass of CH_4 = 16, the molecular mass of 2kmoles of O_2 = 2 x 32 = 64, the molecular mass of CO_2 = 44, while the molecular mass of 2kmoles of H_2O = 2 x 18 = 36. These are the masses shown above. The above equation shows that:

16kg of CH_4 requires 64kg of O_2 for complete combustion. Therefore, by simple proportion 20kg (from the question) of CH_4 will require:

$$\frac{20}{16} \times 64 \quad \text{of } O_2$$

= 80kg of O_2

From the question, air supplied = 380kg. But air contains 23.2% by mass of O_2

Therefore by simple proportion, mass of O_2 in air supplied = $\frac{23.2}{100}$ x 380

= 0.232 x 380

= 88.16kg

Hence, % excess air used = % excess O_2 used

% excess air used = $\frac{O_2 \text{ supplied } - O_2 \text{ required}}{O_2 \text{ required}}$ x 100

= $\frac{88.16 - 80}{80}$ x 100

= 10.2%

Therefore, percent excess air supplied is 10.2%

2. During a combustion process 32 litres of ethene is burnt with 540 litres of air to give 50 litres of carbon (IV) oxide and 18 litres of carbon (II) oxide. Calculate the percent excess air supplied.

(Air contains 21% by volume of oxygen)

Solution

The reaction is a complete combustion reaction since excess air was supplied. The volumes given in the question can be used as moles or kmoles.

The balanced equation for the reaction is:

C_2H_4 + $3O_2$ ----------> $2CO_2$ + $2H_2O$
1kmol + 3kmol -------> 2kmol + 2kmol

This balanced equation shows that 1kmol of C_2H_4 requires 3kmol of O_2 for complete combustion. Therefore, by simple proportion, 32 litres of C_2H_4 will require:

$\frac{32}{1}$ x 3 of O_2 (Note that moles are used as volumes in reactions)

= 96 litres of O_2

However, air supplied = 540 litres. But air contains 21% by volume of O_2

Therefore, by simple proportion, volume of O_2 supplied = $\frac{21}{100}$ x 540

$\qquad$ = 0.21 x 540

$\qquad$ = 113.4 litres

Therefore, % excess air = $\frac{O_2 \text{ supplied } - O_2 \text{ required}}{O_2 \text{ required}}$ x 100

$\qquad$ = $\frac{113.4 - 96}{96}$ x 100

$\qquad$ = 18.1%

3. A gas has the following composition by volume:

5% of CO_2, 12% of CO, 3.8% of C_2H_2, 18% of C_3H_8, 55% of H_2, 0.2% of O_2, and 6% of N_2. The gas is burnt to produce a stack gas containing the following composition by volume on a dry basis: 7.0% of CO_2, 7.6% of O_2, and 85.4% of N_2. Calculate the required and actual air supplied.

Solution

The components of the gas that will undergo combustion are CO, C_2H_4, C_3H_8 and H_2. CO_2, O_2 and N_2 will pass directly to the product stream.

Let us use a basis of 100kmoles of feed.

The combustion reactions for each of these four components that will undergo combustion are as given below.

$\qquad$ CO + ½O_2 ----------> CO_2

$\quad$ 12CO + (12 x ½)O_2 ----------> 12CO_2 $\quad$ (12% of CO represent 12kmol)

$\quad$ 12CO + 6O_2 ----------> 12CO_2 ....Equation 1

$\quad$ C_2H_2 + 3O_2 ----------> 2CO_2 + 2H_2O

$\quad$ 3.8C_2H_2 + (3.8 x 3)O_2 ----------> (3.8 x 2)CO_2 + (3.8 x 2)H_2O $\quad$ (from the 3.8% of C_2H_4 in feed)

$3.8C_2H_2 + 11.4O_2 \longrightarrow 7.6CO_2 + 7.6H_2O$ ....Equation 2

$C_3H_8 + 5O_2 \longrightarrow 3CO_2 + 4H_2O$

$18C_3H_8 + (18 \times 5)O_2 \longrightarrow (18 \times 3)CO_2 + (18 \times 4)H_2O$

$18C_3H_8 + 90O_2 \longrightarrow 54CO_2 + 72H_2O$ ....Equation 3

$55H_2 + (55 \times ½)O_2 \longrightarrow 55H_2O$

$55H_2 + 27.5O_2 \longrightarrow 55H_2O$ ....Equation 4

Bringing equations 1 to 4 together gives:

$12CO + 6O_2 \longrightarrow 12CO_2$ ....Equation 1

$3.8C_2H_2 + 11.4O_2 \longrightarrow 7.6CO_2 + 7.6H_2O$ ....Equation 2

$18C_3H_8 + 90O_2 \longrightarrow 54CO_2 + 72H_2O$ ....Equation 3

$55H_2 + 27.5O_2 \longrightarrow 55H_2O$ ....Equation 4

From these four equations the total kmol of O_2 required = 6 + 11.4 + 90 + 27.5

= 134.9 kmols

Total kmol of CO_2 = 12 + 7.6 + 54

= 73.6

But the 5% of CO_2 in the reactant (from the question) did not react but passed directly to the product. So this is added to the value above as follows:

Total kmol of CO_2 in product = 73.6 + 5

= 78.6

Total kmol of H_2O in product = 7.6 + 72 + 55

= 134.6

Recall that air contains 21% by volume of O_2

Therefore the required O_2 of 134.9 kmol represents 21%

Hence, 100% air will give: $\dfrac{100}{21} \times 134.9$

= 642.8kmol of air

Therefore the theoretical air required is 642.8kmol.

The N_2 in air from the feed will pass directly to the product. Hence, let us calculate the actual air supplied by using the kmol of N_2 in the product (this came from the feed of air).

Recall that the kmol of CO_2 in the product = 78.6. This 78.6kmol represent 7% (i.e. % of CO_2 in the product).

Therefore, by simple proportion, 100% product will be = $\frac{100}{7}$ x 78.6

= 1123kmol

Hence total kmol of product is 1123kmol

Since 85.4% of N_2 is present in the product, then kmol of N_2 in product = $\frac{85.4}{100}$ x 1123

= 959kmol

This 959kmol of N_2 contains the 6% of N_2 from the feed that passed directly to the product stream.

Therefore kmol of N_2 from air = 959 - 6

= 953kmol

Recall that air contains 79% by volume of N_2.

Therefore, 79% N_2 represent 953kmol

Hence, 100% air will be = $\frac{100}{79}$ x 953

= 1206kmol

Therefore actual air supplied is 1206kmols of air.

Another method of solving this problem is to use the individual element, C and H_2 in the hydrocarbons directly.

Therefore the C_2H_4 can be broken into its elements as follows.

C_2H_4 ----------> 2C + 2H_2

Thus, $3.8C_2H_4 \longrightarrow (3.8 \times 2)C + (3.8 \times 2)H_2$

$$3.8C_2H_4 \longrightarrow 7.6C + 7.6H_2 \quad \text{...........................Equation 5}$$

Similarly, C_3H_8 can be broken into its component elements as follows.

$$C_3H_8 \longrightarrow 3C + 4H_2$$

Thus, $18C_3H_8 \longrightarrow (18 \times 3)C + (18 \times 4)H_2$

Or, $18C_3H_8 \longrightarrow 54C + 72H_2$ ....Equation 6

Hence the total kmol of C in the hydrocarbons from equations 5 and 6 is:

$7.6 + 54 = 61.6 \text{kmol}$

The total kmol of H_2 from equations 5 and 6 is:

$7.6 + 72 = 79.6 \text{kmols}$

These total kmol of C and H_2 can now be used to write simple combustion equations as follows.

$$C + O_2 \longrightarrow CO_2$$

$$61.6C + 61.6O_2 \longrightarrow 61.6CO_2 \quad \text{.....................Equation 7}$$

For the H_2 we have:

$$H_2 + \tfrac{1}{2}O_2 \longrightarrow H_2O$$
$$79.6H_2 + (79.6 \times \tfrac{1}{2})O_2 \longrightarrow 79.6H_2O$$

$$79.6H_2 + 39.8O_2 \longrightarrow 79.6H_2O \quad \text{.........................Equation 8}$$

Recall that equations 1 and 4 are:

$$12CO + 6O_2 \longrightarrow 12CO_2 \quad \text{..........................Equation 1}$$

$$55H_2 + 27.5O_2 \longrightarrow 55H_2O \quad \text{.........................Equation 4}$$

Hence, by using equations 7, 8, 1 and 4, we have:

Total kmol of O_2 required = $61.6 + 39.8 + 6 + 27.5$

$= 134.9 \text{kmols}$

Total kmol of CO_2 in product = $61.6 + 12 + 5$ (i.e. CO_2 from feed)

$$= 78.6 \text{kmol of } CO_2$$

Total kmol of H_2O in product = 79.6 + 55

$$= 134.6 \text{ kmol of } H_2O$$

Therefore, these three values of 134.9kmol of O_2, 78.6 kmol of CO_2 and 134.6kmol of H_2O which are the same as obtained in the first method, can now be used to calculate the required and actual air supplied as carried out in the first method.

4. A fuel which contains carbon and hydrogen only, was burnt to produce a dry flue gas with the following composition by volume: 9.6% of CO_2, 8.4% of O_2 and 82.0% of N_2. Calculate:

(a). The composition of the fuel

(b). The percent excess air supplied

Solution

(a). A basis of 100moles of flue gas (product) will be used. This means that the volume of CO_2 produced is 9.6moles, the volume of O_2 produced is 8.4moles, while the volume of N_2 in the product is 82moles. So, we will work from the product back to the reactant.

The combustion reaction for carbon is:

$$C + O_2 \longrightarrow CO_2$$

Since 9.6moles of CO_2 was produced, by working backward, the moles of C and O_2 are obtained as follows:

$$9.6C + 9.6O_2 \longrightarrow 9.6CO_2 \quad \ldots\ldots\ldots\ldots\ldots\text{Equation 1}$$

The N_2 did not undergo any reaction. So, all the 82moles of N_2 came from the air. Recall that air contains 79% by volume of N_2. So by simple proportion:

79% N_2 gives 82moles

Therefore, 100% air will give: $\dfrac{100}{72} \times 82$

= 113.8moles

Therefore, the moles of air supplied is 113.8moles

But air contains 21% by volume of O_2. Therefore, total O_2 supplied from the air is:

$$\frac{21}{100} \times 113.8$$

$$= 0.21 \times 113.8$$

$$= 23.9 \text{moles}$$

From the question 8.4moles of O_2 was present in the product stream. This is the amount of excess O_2 that did not take part in the reaction.

Therefore moles of O_2 that reacted = 23.9 - 8.4

$$= 15.5 \text{moles}$$

Out of this 15.5moles, 9.6moles from equation 1 has reacted with C. Therefore, the remaining O_2 will react with H_2.

Therefore O_2 that react with H_2 = 15.5 - 9.6

$$= 5.9 \text{moles}$$

Hence the H_2 combustion reaction is represented as:

$$H_2 + 5.9O_2 \longrightarrow H_2O$$

In order to balance this reaction, we have to make the oxygen atoms on both sides of the reaction to be 2 x 5.9 (since this is total O_2 atoms on the left). This gives:

$$(2 \times 5.9)H_2 + 5.9O_2 \longrightarrow (2 \times 5.9)H_2O$$

Or, $11.8H_2 + 5.9O_2 \longrightarrow 11.8H_2O$ ....Equation 2

Therefore, from equation 1, the moles of C = 9.6moles, and from equation 2 the moles of H_2 = 11.8moles.

Total moles of C and H_2 = 9.6 + 11.8

$$= 21.4 \text{moles}$$

Hence, the composition of C and H_2 in the fuel, are:

% composition of C = $\frac{9.6}{21.4} \times 100$

$$= 44.9\%$$

% composition of H_2 = $\dfrac{11.8}{21.4}$ x 100

= 55.1%

(b). Recall that the total O_2 supplied = 23.9 moles, and the unreacted (excess) O_2 = 8.4 moles.

Therefore, the reacted (used) O_2 = 23.9 - 8.4

= 15.5 moles

Therefore, % excess air supplied = $\dfrac{\text{supplied } O_2 - \text{required } O_2}{\text{required } O_2}$ x 100

= $\dfrac{\text{excess } O_2}{\text{required } O_2}$ x 100

= $\dfrac{8.4}{15.5}$ x 100

= 54.2%

5. A petroleum gas was analyzed to contain the following composition by volume: 74% of C_3H_8, 22.6% of C_4H_{10}, and 3.4% of CO_2. The gas was burnt with 28% excess air. 92% of the hydrocarbons is converted to CO_2 while 8% is converted to CO. Calculate the composition of the flue gas.

Solution

Let us work on a basis of 100 moles of the petroleum gas. Note that when hydrocarbons burn, the products formed from complete combustion are CO_2 and H_2O. So, the product stream for this equation will contain CO_2, H_2O, CO (from incomplete combustion), O_2 (from excess air) and N_2 (from air).

For complete combustion of each of the hydrocarbons we have:

$C_3H_8 + 5O_2 \longrightarrow 3CO_2 + 4H_2O$

Using the 74 moles (i.e. 74% of C_3H_8), gives:

$74C_3H_8 + (74 \times 5)O_2 \longrightarrow (74 \times 3)CO_2 + (74 \times 4)H_2O$

Or, $74C_3H_8 + 370O_2 \longrightarrow 222CO_2 + 296H_2O$ ....Equation 1

And, $C_4H_{10} + \dfrac{13}{2}O_2 \longrightarrow 4CO_2 + 5H_2O$

Using the 22.6 moles of C_4H_{10} gives:

$$22.6C_4H_{10} + (22.6 \times \frac{13}{2})O_2 \longrightarrow (22.6 \times 4)CO_2 + (22.6 \times 5)H_2O$$

Or, $22.6C_4H_{10} + 146.9O_2 \longrightarrow 90.4CO_2 + 113H_2O$ ....Equation 2

Therefore the total O_2 for complete combustion from equations 1 and 2 is:

$370 + 146.9 = 516.9$ moles

Since 28% excess air was supplied, then the total O_2 supplied is given by:

$\frac{128}{100} \times 516.9$ (28% of excess air means 128% (i.e. 100 + 28) of required air)

$= 1.28 \times 516.9$

$= 661.63$ moles

This amount of O_2 corresponds to 21% by volume of O_2 in air. Therefore, the 79% by volume of N_2 associated with this O_2 will be obtained as follows:

21% of O_2 gives 661.63 moles

Therefore, 79% of N_2 will give: $\frac{79}{21} \times 661.63$

$= 2489.0$ moles of N_2

From the question, the hydrocarbons are also converted to CO as follows:

$$C_3H_8 + \frac{7}{2}O_2 \longrightarrow 3CO + 4H_2O$$

Using the 74 moles of C_3H_8 gives:

$$74C_3H_8 + (74 \times \frac{7}{2})O_2 \longrightarrow (74 \times 3)CO + (74 \times 4)H_2O$$

Or, $74C_3H_8 + 259O_2 \longrightarrow 222CO + 296H_2O$ ....Equation 3

And, $C_4H_{10} + \frac{9}{2}O_2 \longrightarrow 4CO + 5H_2O$

Using the 22.6 moles of C_4H_{10} gives:

$$22.6C_4H_{10} + (22.6 \times \frac{9}{2})O_2 \longrightarrow (22.6 \times 4)CO + (22.6 \times 5)H_2O$$

Or, $22.6C_4H_{10} + 101.7O_2 \longrightarrow 90.4CO + 113H_2O$ ....Equation 4

However, using equations 1 and 2, 92% conversion of the hydrocarbons to CO_2 gives:

$0.92(74C_3H_8 + 370O_2 \text{----------> } 222CO_2 + 296H_2O)$

Or, $68.08C_3H_8 + 340.4O_2 \text{----------> } 204.24CO_2 + 272.32H_2O$ ....Equation 5

And, $0.92(22.6C_4H_{10} + 146.9O_2 \text{----------> } 90.4CO_2 + 113H_2O)$

Or, $20.79C_4H_{10} + 135.15O_2 \text{----------> } 83.17CO_2 + 103.96H_2O$ ....Equation 6

Similarly, from equations 3 and 4, 8% conversion of hydrocarbon to CO gives:

$0.08(74C_3H_8 + 259O_2 \text{----------> } 222CO + 296H_2O)$

Or, $5.92C_3H_8 + 20.72O_2 \text{----------> } 17.76CO + 23.68H_2O$ ....Equation 7

And, $0.08(22.6C_4H_{10} + 101.7O_2 \text{----------> } 90.4CO + 113H_2O)$

Or, $1.81C_4H_{10} + 8.14O_2 \text{----------> } 7.26CO + 9.04H_2O$ ....Equation 8

From equations 5 to 8, the total O_2 used is:

340.4 + 135.15 + 20.72 + 8.14 = 504.41moles

Total CO_2 produced from equations 5 and 6 is:

204.24 + 83.18 = 287.41moles

But total CO_2 in product = 287.41 + 3.4 (i.e. CO_2 from petroleum gas is 3.4)

= 290.81moles

Total H_2O from equations 1 and 2 (i.e. complete combustion) is:

296 + 113 = 409moles

This can also be obtained from equations 5 to 8.

Total CO from equations 7 and 8 is:

17.76 + 7.26 = 25.02moles

Total N_2 in product remains 2489moles

Total O_2 in product (i.e. unreacted or excess O_2) is:

O_2 from excess air supplied - O_2 that reacted

= 661.63 - 504.41

= 157.22 moles

Therefore, total moles of all gases in the product, is:

157.22 moles of O_2 + 2489 moles of N_2 + 25.02 moles of CO + 409 moles of H_2O + 290.81 moles of CO_2 = 3371.05 moles

Therefore the composition of the flue gas is:

$$\% O_2 = \frac{157.22}{3371.05} \times 100$$

= 4.66%

$$\% N_2 = \frac{2489}{3371.05} \times 100$$

= 73.83%

$$\% CO = \frac{25.02}{3371.05} \times 100$$

= 0.74%

$$\% H_2O = \frac{409}{3371.05} \times 100$$

= 12.13%

$$\% CO_2 = \frac{290.81}{3371.05} \times 100$$

= 8.63%

6. A crude oil was analyzed to contain the following composition by mass: 75% of C, 19.4% of H_2 and 5.6% of S. If the stack gas is produced at 250°C and a pressure of 1.04×10^5 N/m^2, calculate:

(a). The composition of the stack gas from combustion of 100kg of the crude oil

(b). The volume of the stack gas.

<u>Solution</u>

(a). Note that the stack gas will contain CO_2 from C, H_2O from H_2, SO_2 from S and N_2 from air used for the combustion.

The complete combustion reaction for C is:

$$C + O_2 \longrightarrow CO_2$$

Using the molecular mass of each substance shows that:

$$12 kgC + 32 kgO_2 \longrightarrow 44 kgCO_2$$

By dividing each substance by 12kg, then 1kg of C will react as follows:

$$1kgC + \frac{32}{12}kgO_2 \longrightarrow \frac{44}{12}kgCO_2$$

Therefore 75kg of C (from question) will react as follows:

$$(75 \times 1)C + (75 \times \frac{32}{12})O_2 \longrightarrow (75 \times \frac{44}{12})CO_2$$

Or, $75C + 200O_2 \longrightarrow 275CO_2$ ....Equation 1

The reaction of H_2 is given by:

$$H_2 + \tfrac{1}{2}O_2 \longrightarrow H_2O$$

Hence, $2kgH_2 + (32 \times \tfrac{1}{2})kgO_2 \longrightarrow 18kgH_2O$

Or, $2kgH_2 + 16kgO_2 \longrightarrow 18kgH_2O$

Hence, by dividing each substance by 2, it means that 1kg of H_2 will react as follows:

$$1kgH_2 + \frac{16}{2}kgO_2 \longrightarrow \frac{18}{2}kgH_2O$$

Or, $1kgH_2 + 8kgO_2 \longrightarrow 9kgH_2O$

Therefore the 19.4kg of H_2 in the crude oil will react as follows:

$$(19.4 \times 1)H_2 + (19.4 \times 8)O_2 \longrightarrow (19.4 \times 9)H_2O$$

Or, $19.4H_2 + 155.2O_2 \longrightarrow 174.6H_2O$ ....Equation 2

The reaction for S is given by:

$$S + O_2 \longrightarrow SO_2$$

$$32 \text{kgS} + 32 \text{kgO}_2 \text{---------->} 64 \text{kgSO}_2 \qquad \text{(Note that the molecular mass of S is 32)}$$

Hence by dividing each substance by 32, it means that 1kg of S will react as follows:

$$1 \text{kgS} + \frac{32}{32} \text{kgO}_2 \text{---------->} \frac{64}{32} \text{kgSO}_2$$

Or, $1 \text{kgS} + 1 \text{kgO}_2 \text{---------->} 2 \text{kgSO}_2$

Therefore the 5.6kg of S in the crude oil will react as follows:

$$(5.6 \times 1)S + (5.6 \times 1)O_2 \text{----------->} (5.6 \times 2)SO_2$$

Or, $5.6S + 5.6O_2 \text{---------->} 11.2SO_2$ ....Equation 3

Therefore from equation 1 to 3, the mass of O_2 required for complete combustion is:

200 + 155.2 + 5.6 = 360.8kg

Recall that air contains 23.2% by mass of O_2. So, by simple proportion, the mass of N_2 in the air can be obtained as follows, and noting that air contains 76.8% by mass of N_2.

23.2% gives 360.8kg of O_2

Therefore, 76.8% will give: $\frac{76.8}{23.2}$ x 360.8 of N_2

= 1194.4kg of N_2

Therefore, mass of N_2 in product is 1194.4kg

Mass of CO_2 in product (from equation 1) is 275kg

Mass of H_2O in product from (from equation 2) is 174.6kg

Mass of SO_2 in product (from equation 3) is 11.2kg

Hence the total mass of the components in the product is:

1194.4 + 275 + 174.6 + 11.2 = 1655.2kg

Therefore, the composition of the stack gas is:

% of $N_2 = \frac{1194.4}{1655.2}$ x 100

= 72.2%

$$\% \text{ of } CO_2 = \frac{275}{1655.2} \times 100$$

$$= 16.6\%$$

$$\% \text{ of } H_2O = \frac{174.6}{1655.2} \times 100$$

$$= 10.5\%$$

$$\% \text{ of } SO_2 = \frac{11.2}{1655.2} \times 100$$

$$= 0.7\%$$

(b). Converting each of the components of the product to kmols gives the following, and noting that:

$$\text{Number of moles} = \frac{\text{mass}}{\text{molecular mass}}$$

Therefore, Kmols of $N_2 = \frac{1194.4}{28}$

$$= 42.68 \text{ kmols}$$

Kmols of $CO_2 = \frac{275}{44}$

$$= 6.25 \text{ kmols}$$

kmols of $H_2O = \frac{174.6}{18}$

$$= 9.7 \text{ kmols}$$

Kmols of $SO_2 = \frac{11.2}{64}$

$$= 0.18$$

Hence, total kmols of stack gas = 42.68 + 6.25 + 9.7 + 0.18

$$= 58.81 \text{ kmols}$$

Recall that 1kmol of any gas occupies a volume of 22.4m³. Therefore the volume occupied by 58.81kmols of stack gas is:

58.81 x 22.4 = 1317.3m³

Therefore, by using the general gas law, the initial conditions of the stack gas at s.t.p are:

$P_1 = 1.01 \times 10^5 N/m^2$, $T_1 = 273k$, $V_1 = 1317.3m^3$. The final conditions of the stack gas are:

$P_2 = 1.04 \times 10^5 N/m^2$, $T_2 = 250 + 273 = 523k$, $V_2 = ?$ (Note that 250°C is given in the question)

Hence, from the general gas law:

$$\frac{P_1V_1}{T_1} = \frac{P_2V_2}{T_2}$$

Therefore, $V_2 = \dfrac{P_1V_1T_2}{P_2T_1}$

$= \dfrac{1.01 \times 1317.3 \times 523}{1.04 \times 273}$ (Note that 10^5 from pressure has canceled out)

$V_2 = 2450.8 m^3$

7. Ethane is burnt with 20% excess air. Fractional conversion is 84%. Calculate the percentage composition of each of the components of the flue gas.

Solution

For complete combustion the equation below follows:

$$C_2H_6 + \frac{7}{2}O_2 \longrightarrow 2CO_2 + 3H_2O$$

Working on a basis of 1mole of ethane as the feed, the equation above shows that:

$\dfrac{7}{2}$ or 3.5moles of O_2 is required

But, 20% excess air was supplied.

Therefore, moles of O_2 supplied $= \dfrac{120}{100} \times 3.5$

$= 1.2 \times 3.5$

$= 4.2$ moles

Air contains 21% O_2 and 79% N_2 by volume. So, the moles of N_2 in the air supplied, is obtained as follows:

21% O_2 gives 4.2moles

Therefore, 79% will give: $\frac{79}{21} \times 4.2$

$= 15.8$ moles of N_2

This shows that the moles of N_2 in the product is 15.8moles

Now, for the 84% fractional conversion, the above reaction equation is given as:

$$0.84(C_2H_6 + 3.5O_2 \text{----------->} 2CO_2 + 3H_2O)$$

Or, $\quad 0.84C_2H_6 + 2.94O_2 \text{---------->} 1.68CO_2 + 2.52H_2O$ ....Equation 1

It follows from this reaction that 0.84 moles of C_2H_6 reacted. So, moles of C_2H_6 that is in product (i.e. unreacted C_2H_6) is given by:

$= 1 - 0.84 \quad$ (Note that feed is 1 mole of C_2H_6)

$= 0.16$ moles

From equation 1 above, 2.94 moles of O_2 reacted. So moles of O_2 that is in the product (i.e. unreacted O_2) = 4.2 - 2.94

$= 1.26$ moles

Moles of CO_2 produced, i.e. CO_2 in product = 1.68 moles

Moles of H_2O in product = 2.52 moles   (All from equation 1)

Thus the flue gas contains:

15.8 moles of N_2

0.16 moles of C_2H_6

1.26 moles of O_2

1.68 moles of CO_2

2.52 moles of H_2O

Total moles = 21.42   (When all the values above are added)

Therefore the flue gas contains the following percentage composition:

% $N_2 = \frac{15.8}{21.42} \times 100$

$$= 73.8\%$$

$$\% \, C_2H_6 = \frac{0.16}{21.42} \times 100$$

$$= 0.7\%$$

$$\% \, O_2 = \frac{1.26}{21.42} \times 100$$

$$= 5.9\%$$

$$\% \, CO_2 = \frac{1.68}{21.42} \times 100$$

$$= 7.8\%$$

$$\% \, H_2O = \frac{2.52}{21.42} \times 100$$

$$= 11.8\%$$

EXERCISE

1. 60kg of methane is combusted with 1100kg of air to give 85kg of carbon (IV) oxide and 42kg of carbon (II) oxide. Determine the percent excess air used.

(Air contains 23.2% by mass of oxygen, and the molecular mass of air is 29kg/kmol)

2. During a combustion process 250 litres of ethene is burnt with 4800 litres of air to give 440 litres of carbon (IV) oxide and 120 litres of carbon (II) oxide. Calculate the percent excess air supplied.
(Air contains 21% by volume of oxygen)

3. A gas has the following composition by volume:
7% of CO_2, 15% of CO, 2.4% of C_2H_2, 16.5% of C_3H_8, 48% of H_2, 0.8% of O_2, and 10.3% of N_2. The gas is burnt to produce a stack gas containing the following composition by volume on a dry basis: 8.5.0% of CO_2, 6.9% of O_2, and 84.6% of N_2. Calculate the required an actual air supplied.

4. A fuel which contains carbon and hydrogen only, was burnt to produce a dry flue gas with the following composition by volume: 13.2% of CO_2, 7.1% of O_2 and 79.7% of N_2. Calculate:

(a). The composition of the fuel

(b). The percent excess air supplied

5. A petroleum gas was analyzed to contain the following composition by volume: 80% of C_2H_6, 15.2% of C_4H_{10}, and 4.8% of CO_2. The gas was burnt with 32% excess air. 90% of the hydrocarbons is converted to CO_2 while 10% is converted to CO. Calculate the composition of the flue gas.

6. A crude oil was analyzed to contain the following composition by mass: 84% of C, 12.4% of H_2 and 3.6% of S. If the stack gas is produced at 220°C and a pressure of 1.08×10^5 N/m², calculate:

(a). The composition of the stack gas from combustion of 100kg of the crude oil

(b). The volume of the stack gas.

7. Propane is burnt with 10% excess air. Fractional conversion is 78%. Calculate the percentage composition of each of the components of the flue gas.

CHAPTER 4
CALCULATIONS INVOLVING LIMITING REACTANTS

A limiting reactant is the component of a chemical reaction which is available for complete conversion in the reaction. It limits the excess reactant from reacting completely in the reaction.
An excess reactant is a reactant which is not completely used up in a reaction.
The percentage of the limiting reactant which actually reacts in a reaction is known as the degree of completion of the reaction.

Examples

1. Iron (II) sulphide was produced by heating five parts of iron with four parts of sulphur to produce 70% of iron (II) sulphide. The converter has a capacity of 1000kg. Determine:

(a). the limiting reactant

(b). the percent excess reactant

(c). the degree of completion of the reaction

Solution

(a). Let us use a basis of 1000kg of reactants.

The equation for the reaction is:

$$Fe + S \longrightarrow FeS$$

Five parts of Fe and four parts of S means that the ratio of Fe to S is 5 : 4

The total ration = 5 + 4 = 9

Therefore mass of Fe in feed = $\frac{5}{9}$ x 1000

= 555.6kg

The mass of S in feed = $\frac{4}{9}$ x 1000

= 444.4kg

Assuming that: Input = Output, then:

70% of FeS formed in the product means that:

Mass of FeS in product = $\dfrac{70}{100} \times 1000$

$= 700\text{kg}$

Converting these masses to kmols gives:

kmols of Fe = $\dfrac{555.6}{56}$ (Note that the molecular mass of Fe = 56)

$= 9.9\text{kmols}$

kmols of S = $\dfrac{444.4}{32}$ (Note that the relative molecular mass of S = 32)

$= 13.9\text{kmols}$

kmols of FeS = $\dfrac{700}{56+32} = \dfrac{700}{88}$

$= 8.0\text{kmols}$

The reaction is:

Fe + S ----------> FeS

The reaction shows that 1kmol of Fe reacts with 1kmol of S to produce 1kmol of FeS

Therefore, 8kmols of FeS will be produced from:

8Fe + 8S ----------> 8FeS

8kmols of FeS will be produced from 8kmols of Fe and 8kmols of S.

However, 9.9kmols of Fe supplied would also require 9.9kmols of S and not 13.9kmols of S. Hence S is present in excess.

This means that Fe is present in limited amount. Hence it is the limiting reactant.

(b). The excess reactant is S. 9.9kmols of S is required. But 13.9kmols is supplied. Therefore excess amount of S supplied = 13.9 - 9.9 = 4kmols

Therefore, % excess reactant = $\dfrac{\text{excess kmols}}{\text{required kmols}} \times 100$

$= \dfrac{4}{9.9} \times 100$

$= 40.4\%$

(c). Fe is the limiting reactant. The amount of Fe supplied is 9.9kmols.

The amount that reacted is 8kmols

Therefore, the degree of completion of the reaction is = $\frac{8}{9.9}$ x 100

= 80.8%

2. Calcium oxide is produced by the burning of calcium in oxygen. If 100kmols of calcium is mixed with 100kmols of oxygen, determine:

(a). the limiting reactant

(b). the percent excess of the excess reactant

(c). the composition of the product stream based on a complete combustion

(d). The composition of the product stream based on a 70% fractional conversion of the limiting reactant.

Solution

(a). The reaction is as shown below:

$2Ca + O_2 \longrightarrow 2CaO$

The reaction shows that 2kmols of Ca reacts with 1kmol of O_2 to produce 2kmols of CaO.

Therefore, 100kmols of Ca will require 50kmols of O_2.

But, 100kmols of O_2 is available. This shows that O_2 is present in excess.

Hence Ca is the limiting reactant.

(b). Of the 100kmols of O_2, only 50kmols will react with Ca.

Therefore the excess O_2 = 100 - 50

= 50kmols

Percent excess of the excess reactant (i.e. O_2) = $\frac{excess}{required}$ x 100

= $\frac{50}{50}$ x 100

$$= 100\%$$

(c). All the 100kmols of Ca will react with 50kmols of O_2 to produce 100kmols of CaO. Therefore the product will contain 50kmols of O_2 (unreacted O_2) and 100kmols of CaO.

Therefore, total kmols of product = 50 + 100

$$= 150 \text{kmols}$$

% of O_2 in product $= \dfrac{50}{150} \times 100$

$$= 33.3\%$$

% of CaO in product $= \dfrac{100}{150} \times 100$

$$= 66.7\%$$

(d). The reaction in terms of kmols is as follows:

$$100Ca + 50O_2 \longrightarrow 100CaO$$

A 70% fractional conversion will give:

$$0.7(100Ca + 50O_2 \longrightarrow 100CaO)$$

Or, $\quad 70Ca + 35O_2 \longrightarrow 70CaO$

This means that:

Unreacted Ca = 100 - 70

$$= 30 \text{kmols}$$

Unreacted O_2 = 100 - 35

$$= 65 \text{kmols}$$

Therefore the product stream contains:

70kmols of CaO (i.e. product formed)

30kmols of Ca (unreacted Ca)

65kmols of O_2 (unreacted O_2)

Hence total kmols in product stream = 70 + 30 + 65

$$= 165 \text{ kmols}$$

Therefore, % CaO in product $= \dfrac{70}{165} \times 100$

$$= 42.4\%$$

% of Ca in product $= \dfrac{30}{165} \times 100$

$$= 18.2\%$$

% of O_2 in product $= \dfrac{65}{165} \times 100$

$$= 39.4\%$$

3. An ore consist of pure $MgSO_4$. It is fused with carbon. The composition of the fusion mass is:

    8.5% of $MgSO_4$

    76.4% of MgS

    15.1% of C

The equation for the reaction is:

$$MgSO_4 + 4C \longrightarrow MgS + 4CO$$

Determine:

(a). the excess reactant

(b). percent excess reactant

(c). degree of completion of the reaction

Solution

(a). The molecular mass of $MgSO_4$ = 24 + 32 + (16 x 4)

$$= 24 + 32 + 64$$

$$= 120$$

The molecular mass of MgS = 24 + 32

$$= 56$$

The molecular mass of CO = 12 + 16

$$= 28$$

Therefore the equation can now be expressed along with the molecular mass as follows:

$$MgSO_4 + 4C \longrightarrow MgS + 4CO$$
$$120 + (4 \times 12) \longrightarrow 56 + (4 \times 28)$$
Or, $$120 + 48 \longrightarrow 56 + 112$$

This can be written as

$$120 kgMgSO_4 + 48 kgC \longrightarrow 56 kgMgS + 112 kgCO$$

Using MgS as the major product, 1 kg of MgS will be produced by dividing each substance by 56 (i.e. the molecular mass of MgS) as follows:

$$\frac{120}{56} kgMgSO_4 + \frac{48}{56} kgC \longrightarrow \frac{56}{56} kgMgS + \frac{112}{56} kgCO$$

Or,   1kgMgS is obtained from: $\frac{120}{56} kgMgSO_4 + \frac{48}{56} kgC$

Working on a basis of 100kg of the product, 76.4kg of MgS (i.e. composition from the question) is obtained from the reactants as follows:

$$76.4 kgMgS \text{ is obtained from: } 76.4\left(\frac{120}{56} kgMgSO_4 + \frac{48}{56} kgC\right)$$

$$= 163.7 kgMgSO_4 + 65.5 kgC \quad \text{(obtained after simplifying the bracket above)}$$

This shows that in order to obtain 76.4kg of MgS, the reactants will be:

163.7kg of $MgSO_4$ and 65.5kg of C.

Since the product composition shows that 8.5kg of $MgSO_4$ is present in the product, then the total mass of $MgSO_4$ supplied = Reacted $MgSO_4$ + $MgSO_4$ in product (i.e. unreacted $MgSO_4$)

Therefore, $MgSO_4$ supplied = 163.7 + 8.5 = 172.2kg

Similarly, C supplied = 65.5 + 15.1

$$= 80.6 kg$$

Expressing these masses in kmols gives:

$$\text{kmols of MgSO}_4 \text{ supplied} = \frac{172.2}{120}$$

$$= 1.44 \text{kmols}$$

$$\text{kmols of C supplied} = \frac{80.6}{12}$$

$$= 6.72 \text{kmols}$$

Recall that the reaction is given by:

$$MgSO_4 + 4C \longrightarrow MgS + 4CO$$

This shows that 1kmol of $MgSO_4$ requires 4kmols of C

Therefore, 1.44kmols of $MgSO_4$ will require: $\frac{1.44}{1} \times 4$

$$= 5.76 \text{kmols of C}$$

However, the C supplied is 6.72kmols. This is above the required value of 5.76kmols. Hence, C is in excess.

Therefore the excess reactant is C.

(b). Kmols C supplied = 6.72kmols

Kmols C required = 5.76kmols

Therefore, % excess reactant = $\frac{\text{supplied} - \text{required}}{\text{required}} \times 100$

$$= \frac{6.72 - 5.76}{5.76} \times 100$$

$$= 16.7\%$$

(c). $MgSO_4$ is the limiting reactant. The amount of $MgSO_4$ supplied is 172.2kg. The amount that reacted is 163.7kg

Therefore, the degree of completion of the reaction is = $\frac{\text{Amount of MgSO}_4 \text{ reacted}}{\text{Amount of MgSO}_4 \text{ supplied}} \times 100$

$$= \frac{163.7}{172.2} \times 100$$

$$= 95.1\%$$

4. Iron (II) chloride was produced by heating 3 parts of iron with 2 parts of chlorine to produce 62% of iron (II) chloride. The converter has a capacity of 500kg. Determine:

(a). the limiting reactant

(b). the percent excess reactant

(c). the degree of completion of the reaction

Solution

(a). Let us use a basis of 500kg of reactants.

The equation for the reaction is:

$$Fe + Cl_2 \longrightarrow FeCl_2$$

3 parts of Fe and 2 parts of Cl_2 means that the ratio of Fe to Cl_2 is 3 : 2

The total ration = 3 + 2 = 5

Therefore mass of Fe in feed = $\frac{3}{5}$ x 500

= 300kg

The mass of Cl_2 in feed = $\frac{2}{5}$ x 500

= 200kg

Assuming that: Input = Output, then:

62% of $FeCl_2$ produced means that:

Mass of $FeCl_2$ produced = $\frac{62}{100}$ x 500

= 310kg

Converting these masses to kmols gives:

kmols of Fe = $\frac{300}{56}$ (Note that the molecular mass of Fe = 56)

= 5.4kmols

kmols of $Cl_2 = \dfrac{200}{71}$ (Note that the relative molecular mass of $Cl_2 = 2 \times 35.5 = 71$)

$\qquad\qquad\quad = 2.8 \text{kmols}$

kmols of $FeCl_2 = \dfrac{310}{56+71}$

$\qquad\qquad\quad = \dfrac{310}{127}$

$\qquad\qquad\quad = 2.4 \text{kmols}$

The reaction is:

$\qquad$ Fe + Cl_2 ----------> $FeCl_2$

The reaction shows that 1kmol of Fe reacts with 1kmol of Cl_2 to produce 1kmol of $FeCl_2$

Therefore, 2.4kmols of $FeCl_2$ will be produced from:

$\qquad$ 2.4Fe + 2.4Cl_2 ----------> 2.4$FeCl_2$

2.4kmols of $FeCl_2$ will be produced from 2.4kmols of Fe and 2.4kmols of Cl_2.

However, 5.4kmols of Fe supplied would also require 5.4kmols of Cl_2 and not 2.8kmols of Cl_2. Hence Fe is present in excess.

This means that Cl_2 is present in limited amount. Hence, it is the limiting reactant.

(b). The excess reactant is Fe. 2.4kmols of Fe is required. But 5.4kmols is supplied. Therefore excess amount of Fe supplied = 5.4 - 2.4 = 3kmols

Therefore, % excess reactant = $\dfrac{\text{excess kmols}}{\text{required kmols}} \times 100$

$\qquad\qquad\qquad\qquad\quad = \dfrac{3}{2.4} \times 100$

$\qquad\qquad\qquad\qquad\quad = 125\%$

(c). Cl_2 is the limiting reactant. The amount of Cl_2 supplied is 2.8kmols. The amount that reacted is 2.4kmols.

Therefore, the degree of completion of the reaction is = $\dfrac{\text{Amount of } Cl_2 \text{ reacted}}{\text{Amount of } Cl_2 \text{ supplied}} \times 100$

$\qquad\qquad\qquad\qquad = \dfrac{2.4}{2.8} \times 100$

= 85.7%

5. A tin stone consists of pure SnO_2. It is fused with charcoal. The composition of the fusion mass is:

6.8% of SnO_2

82% of Sn

11.2% of C

The equation for the reaction is:

$SnO_2 + 2C \longrightarrow Sn + 2CO$

Determine:

(a). the excess reactant

(b). percent excess reactant

(c). degree of completion of the reaction

Solution

(a). The molecular mass of SnO_2 = 119 + (16 x 2)   (Note that the atomic mass of Sn = 119)

= 119 + 32

= 151

The atomic mass of Sn = 119

The molecular mass of CO = 12 + 16

= 28

Therefore the equation can now be expressed along with the molecular mass as follows:

$SnO_2 + 2C \longrightarrow Sn + 2CO$
151 + (2 x 12) -------> 119 + (2 x 28)
Or,   151 + 24 ----------> 119 + 56

This can be written as

151kgSnO_2 + 24kgC ----------> 119kgSn + 56kgCO

Using Sn as the major product, 1kg of Sn will be produced by dividing each substance by 119 (i.e. the mass of Sn) as follows:

$$\frac{151}{119}kgSnO_2 + \frac{24}{119}kgC \longrightarrow \frac{119}{119}kgSn + \frac{56}{119}kgCO$$

Or, 1kgSn is obtained from: $\frac{151}{119}kgSnO_2 + \frac{24}{119}kgC$

Working on a basis of 100kg of the product, 82kg of Sn (i.e. composition from the question) is obtained from the reactants as follows:

$$82kgSn \text{ is obtained from: } 82(\frac{151}{119}kgSnO_2 + \frac{24}{119}kgC)$$

= 104kgSnO$_2$ + 16.5kgC   (obtained after simplifying the bracket above)

This shows that in order to obtain 82kg of Sn, the reactants will be:

104kg of SnO$_2$ and 16.5kg of C.

Since the product composition shows that 6.8kg of SnO$_2$ is present in the product, then the total mass of SnO$_2$ supplied = Reacted SnO$_2$ + SnO$_2$ in product (i.e. unreacted SnO$_2$)

Therefore, SnO$_2$ supplied = 104 + 6.8 = 110.8kg

Similarly, C supplied = 16.5 + 11.2

= 27.7kg

Expressing these masses in kmols gives:

$$\text{kmols of SnO}_2 \text{ supplied} = \frac{110.8}{151}$$

= 0.73kmols

$$\text{kmols of C supplied} = \frac{27.7}{12}$$

= 2.3kmols

Recall that the reaction is given by:

SnO$_2$ + 2C ----------> Sn + 2CO

This shows that 1kmol of SnO$_2$ requires 2kmols of C

Therefore, 0.73kmols of SnO$_2$ will require: $\frac{0.73}{1}$ x 2

= 1.46kmols of C

However, the C supplied is 2.3kmols. This is above the required value of 1.46kmols. Hence, C is in excess.

Therefore the excess reactant is C.

(b). Kmols C supplied = 2.3kmols

Kmols C required = 1.46kmols

Therefore, % excess reactant = $\frac{\text{supplied} - \text{required}}{\text{required}}$] x 100

$= \frac{2.3 - 1.46}{1.46}$ x 100

= 57.5%

(c). SnO$_2$ is the limiting reactant. The amount of SnO$_2$ supplied is 110.8kg. The amount that reacted is 104kg

Therefore, the degree of completion of the reaction is = (Amount of SnO$_2$ reacted/Amount of SnO$_2$ supplied) $\frac{\text{Amount of SnO}_2 \text{ that reacted}}{\text{Amount of SnO}_2 \text{ supplied}}$ x 100

$= \frac{104}{110.8}$ x 100

= 93.9%

EXERCISE

1. Iron (II) sulphide was produced by heating seven parts of iron with five parts of sulphur to produce 82% of iron (II) sulphide. The converter has a capacity of 2000kg. Determine:

(a). the limiting reactant

(b). the percent excess reactant

(c). the degree of completion of the reaction

2. Calcium oxide is produced by the burning of calcium in oxygen. If 500kmols of calcium is mixed with 200kmols of oxygen, determine:

(a). the limiting reactant

(b). the percent excess of the excess reactant

(c). the composition of the product stream based on a complete combustion

(d). The composition of the product stream based on a 90% fractional conversion of the limiting reactant.

3. An ore consist of pure $MgSO_4$. It is fused with carbon. The composition of the fusion mass is:

7.8% of $MgSO_4$

80% of MgS

12.2% of C

The equation for the reaction is:

$$MgSO_4 + 4C \longrightarrow MgS + 4CO$$

Determine:

(a). the excess reactant

(b). percent excess reactant

(c). degree of completion of the reaction

4. Iron (III) oxide was produced by heating 6 parts of iron with 5 parts of oxygen to produce 75% of iron (III) oxide. The converter has a capacity of 2000kg. Determine:

(a). the limiting reactant

(b). the percent excess reactant

(c). the degree of completion of the reaction

5. A tin stone consists of pure SnO_2. It is fused with charcoal. The composition of the fusion mass is:

6.6% of SnO_2

85% of Sn

8.4% of C

The equation for the reaction is:

$$SnO_2 + C \longrightarrow Sn + CO_2$$

Determine:

(a). the excess reactant

(b). percent excess reactant

(c). degree of completion of the reaction

CHAPTER 5
CALCULATIONS INVOLVING THE FORMULA OF COMPOUNDS

The formula of a compound shows the number of atoms present in one molecule of the compound. This formula can either be empirical formula or molecular formula.

The empirical formula is a formula which shows the ratio of the number of atoms of elements present in a compound.

The molecular formula of a compound shows the exact number of atoms of element present in a compound. For example the molecular formula of benzene is C_6H_6, while its empirical formula is CH, which is a ratio of 1 : 1. Empirical formula can be determined from the constituent masses or percentage composition of the elements in the compound. Molecular formula can be obtained from empirical formula if the molecular mass of the compound is known. The examples given below show the various methods of finding the formulae of compounds.

Examples

1. 4.675kg of an oxide of copper on reduction yielded 4.15 of copper. Calculate the empirical formula of the oxide (Cu = 63.5, O = 16)

Solution

Mass of oxide = 4.675g

Mass of copper = 4.15g

Therefore, mass of oxygen = 4.675 - 4.15

= 0.525g

The calculation of the empirical formula can be set out as follows:

	Cu	O
Mass of element:	4.15	0.525
Moles of element:	$\frac{4.15}{63.5} = 0.0654$	$\frac{0.525}{16} = 0.0328$
Divide by the smallest moles	$\frac{0.0654}{0.0328} = 2$	$\frac{0.0328}{0.0328} = 1$

Therefore the ratio of the atoms is 2 : 1

Hence, the empirical formula is Cu_2O (This means 2 atoms of copper and 1 atom of oxygen)

2. On analysis, 1.24g of an oxide of sulphur was found to contain 0.62g of sulphur. Calculate the:

(a). percentage by mass of each element in the compound

(b). empirical formula of the compound

(S = 32, O = 16)

Solution

(a). % by mass of sulphur = $\frac{0.62}{1.24}$ x 100

$\qquad$ = 50%

% by mass of oxygen = 100 - 50

$\qquad$ = 50%

(b). The empirical formula is calculated as follows:

	S	O
% composition of element:	50%	50%
Ratio of atoms:	$\frac{50}{32}$ = 1.563	$\frac{50}{16}$ = 3.125
Divide each by the smallest ratio	$\frac{1.563}{1.563}$ = 1	$\frac{3.125}{1.563}$ = 2

Therefore, the empirical formula is SO_2. (This means 1 atom of sulphur and 2 atoms of oxygen)

3. The percentage composition of a compound is given by: Sodium 16.2%, Carbon 4.1%, Oxygen 16.9%, and water of crystallization 62.8%. Calculate the formula of the compound. (Na = 23, C = 12, O = 16, H = 1)

Solution

	Na	C	O	H_2O
% Composition	16.2%	4.1%	16.9%	62.8

Ratio of atoms	$\frac{16.2}{23} = 0.704$	$\frac{4.1}{12} = 0.342$	$\frac{16.9}{16} = 1.06$	$\frac{62.8}{18} = 3.49$
Divide by smallest ratio	$\frac{0.704}{0.342} = 2$	$\frac{0.342}{0.342} = 1$	$\frac{1.06}{0.342} = 3$	$\frac{3.49}{0.342} = 10$

This shows that the compound contains 2 atoms of Na, 1 atom of C, 3 atoms of O and 10 molecules of H_2O.

Hence the formula of the compound is $Na_2CO_3.10H_2O$

4. $8cm^3$ of a hydrocarbon is put in a eudiometer tube and sparked with excess oxygen. After the combustion, the volume of the resulting gaseous mixture was found to be $102cm^3$. An introduction of a pellet of sodium hydroxide reduced the volume to $70cm^3$. After cooling the tube to room temperature, the volume of gas left was found to be $30cm^3$. Determine:

(a). the formula of the hydrocarbon

(b). the volume of oxygen sparked with the hydrocarbon

Solutions

From the question the following information can be obtained.

Volume of carbon (IV) oxide = volume of gas absorbed by sodium hydroxide = 102 - 70 = $32cm^3$

Volume of water = volume of gas condensed after cooling (Since the steam condensed to water) = 70 - 30 = $40cm^3$.

Note that a hydrocarbon contains carbon and hydrogen only, and it burns completely in oxygen to produce carbon (IV) oxide and steam (water).

Let the hydrocarbon be C_xH_y. Therefore the reaction can be written as follows.

$C_xH_y + O_2 \longrightarrow xCO_2 + \frac{y}{2}H_2O$ (Note that x and y have been used to balance the equation)

$8cm^3$ ---------> $32cm^3$ $40cm^3$

1Vol.   ---------> 4Vol.   5Vol. (After dividing throughout by 8 to make C_xH_y 1 volume)

1 mole   ---------> 4moles   5moles

Therefore, 1 mole C_xH_y produces ---------> 4 moles CO_2 and 5 moles H_2O

This shows from the equation that x = 4 and $\frac{y}{2} = 5$

If $\frac{y}{2} = 5$, the y = 2 x 5

= 10

Therefore, x = 4 and y = 10.

Hence the hydrocarbon is C_XH_Y which is C_4H_{10} (when x is replaced by 4 and y is replaced by 10).

(b). The balanced equation using 1 mole of the hydrocarbon is given by:

$C_4H_{10} + \frac{13}{2}O_2 \longrightarrow 4CO_2 + 5H_2O$ (Use the total oxygen on the right to balance oxygen on the left)

Hence, 1mole + $\frac{13}{2}$moles ----> 4moles + 5 moles

Multiplying each of the value above by 8cm³ (i.e. volume of the hydrocarbon) gives:

$8cm^3 + 52cm^3 \longrightarrow 32cm^3 + 40cm^3$

This shows that the volume of oxygen that reacted is 52cm³. Recall that the final volume of gas left after cooling the mixture was 30cm³. This is the volume of the unreacted oxygen. Therefore the volume of oxygen sparked with the hydrocarbon is given by:

Reacted oxygen + unreacted oxygen

= 52 + 30

= 82cm³.

Therefore, 82cm³ of oxygen was sparked with the hydrocarbon.

5. 7.2cm³ of a hydrocarbon is sparked with 50.2cm³ of oxygen. After reaction and cooling of the vessel, the volume of the gaseous mixture was found to be 32.2cm³. Sodium hydroxide was added to the reaction vessel. This reduced the gaseous mixture to 3.4cm³. Determine the formula of the hydrocarbon.

Solution

Volume of hydrocarbon = 7.2cm³

Volume of carbon (IV) oxide = 32.2 - 3.4

= 28.8cm³

Volume of unreacted oxygen (i.e. final gas left) = 3.4cm³

Therefore, volume of reacted oxygen = 50.2 - 3.4

$$= 46.8 cm^3$$

Let the hydrocarbon be C_XH_Y. Therefore the reaction can now be written and balanced in terms of x and y as shown below.

$$C_XH_Y + O_2 \longrightarrow xCO_2 + \frac{y}{2}H_2O$$

7.2 + 46.8 ---------> 28.8

1mole + 6.5mole ----> 4moles   (After dividing throughout by 7.2 to make C_XH_Y 1 mole)

The equation can now be written as follows:

$$C_XH_Y + 6\frac{1}{2}O_2 \longrightarrow 4CO_2 + \frac{y}{2}H_2O$$

Or, $C_XH_Y + \frac{13}{2}O_2 \longrightarrow 4CO_2 + 5H_2O$ (After balancing the oxygen atoms)

From the reaction above, there are 13 atoms ($\frac{13}{2}$ x 2 = 13) of oxygen on the left hand side. Therefore in order to balance the oxygen atoms (i.e. to also have 13 oxygen atoms on the right), there has to be 5 moles of H_2O on the right hand side. This is how we got the 5 in the reaction above.

Hence, $\frac{y}{2}H_2O = 5H_2O$

$$\frac{y}{2} = 5$$

Therefore,   y = 2 x 5

y = 10

Similarly, $xCO_2 = 4CO_2$

Therefore,   x = 4

Hence the hydrocarbon, C_XH_Y, is C_4H_{10}.

6. 20cm³ of a hydrocarbon was mixed with 110cm³ of oxygen and exploded. After cooling to room temperature, 80cm³ of gas was left. Addition of potassium hydroxide absorbed 40cm³ of the gas. Determine the formula of the hydrocarbon.

<u>Solution</u>

Volume of hydrocarbon = 20cm^3

Volume of carbon (IV) oxide = 40cm^3

Volume of unreacted oxygen = 80 - 40   (Note that unreacted oxygen is the gas left)

$$= 40cm^3$$

Therefore, volume of reacted oxygen = 110 - 40

$$= 70cm^3$$

The equation of the reaction is now written as follows:

$$C_xH_y + O_2 \longrightarrow xCO_2 + \frac{y}{2}H_2O$$

20   70 ----------> 40

1mole + $3\frac{1}{2}$moles --> 2moles   (After dividing each value by 20 to make C_xH_y 1 mole)

Therefore, $C_xH_y + \frac{7}{2}O_2 \longrightarrow 2CO_2 + 3H_2O$

From the reaction above, there are 7 atoms ($\frac{7}{2}$ x 2 = 7) of oxygen on the left hand side. Therefore in order to balance the oxygen atoms (i.e. to also have 7 oxygen atoms on the right), there has to be 2 moles of CO_2 and 3 moles of H_2O on the right hand side. This is how we got the 2 and 3 in the reaction above.

Hence, by comparing the first and last equations, we see that: $\frac{y}{2}$ = 3, which simplifies to y = 2 x 3 = 6.

Similarly x = 2.

Therefore the hydrocarbon, C_xH_y, is C_2H_6.

7. 12cm^3 of a hydrocarbon is sparked with excess oxygen. After the combustion, the volume of the resulting gaseous mixture was found to be 55.2cm^3. An introduction of a pellet of sodium hydroxide reduced the volume to 43.2cm^3. After cooling the tube to room temperature, the volume of gas left was found to be 19.2cm^3. Determine:

(a). the formula of the hydrocarbon

(b). the volume of oxygen sparked with the hydrocarbon

Solutions

From the question the following information can be obtained.

Volume of carbon (IV) oxide = volume of gas absorbed by sodium hydroxide = 55.2 - 43.2 = 12cm^3

Volume of water = volume of gas condensed after cooling (Since the steam condensed to water) = 43.2 - 19.2 = 24cm^3.

Let the hydrocarbon be C_XH_Y. Therefore the reaction can be written as follows.

$$C_XH_Y + O_2 \longrightarrow xCO_2 + \frac{y}{2}H_2O$$ (Note that x and y have been used to balance the equation)

12cm^3	---------->	12cm^3	24cm^3
1 Vol.	---------->	1 Vol.	2 Vol. (After dividing throughout by 12 to make C_XH_Y 1 volume)
1 mole	---------->	1 moles	2 moles

Therefore, 1 mole C_XH_Y produces ----------> 1 mole CO_2 and 2 moles H_2O

This shows from the equation that x = 1 and $\frac{y}{2}$ = 2

If $\frac{y}{2}$ = 2, the y = 2 x 2

= 4

Therefore, x = 1 and y = 4.

Hence the hydrocarbon, C_XH_Y, is C_1H_4, which is CH_4

(b). The balanced equation using 1 mole of the hydrocarbon is given by:

$CH_4 + 2O_2 \longrightarrow CO_2 + 2H_2O$ (Use the total oxygen on the right to balance oxygen on the left)

1mole + 2moles ----> 1mole + 2 moles

Multiplying each of the value above by 12 (i.e. volume of the hydrocarbon) gives:

12cm^3 + 24cm^3 ----------> 12cm^3 + 24cm^3

This shows that the volume of oxygen that reacted is 24cm^3. Recall that the final volume of gas left after cooling the mixture was 19.2cm^3. This is the volume of the unreacted oxygen. Therefore the volume of oxygen sparked with the hydrocarbon is given by:

Reacted oxygen + unreacted oxygen

= 24 + 19.2

= 43.2cm³.

Therefore, 43.2cm³ of oxygen was sparked with the hydrocarbon.

8. 1.56g of a hydrocarbon produced 5.28g of carbon (IV) oxide and 1.08g of water when burnt in air. Find:

(a). the empirical formula of the hydrocarbon

(b). its molecular formula if its molecular mass is 78

(C = 12, O = 16, H = 1)

Solution

(a). Molecular mass of CO_2 = 44, and atomic mass of carbon, C, is 12.

Therefore the mass of carbon in 5.28g of carbon (IV) oxide is given by simple proportion as follows:

$\frac{12}{44}$ x 5.24 = 1.44g

Similarly, the mass of hydrogen (H_2 = 2) in 1.08g of water (H_2O = 18) is given by:

$\frac{2}{18}$ x 1.08 = 0.12g

Therefore the empirical formula is calculated as follows:

	C	H
Mass of element:	1.44	0.12
Moles of element:	$\frac{1.44}{12}$ = 0.12	$\frac{0.12}{1}$ = 0.12
Divide each by the smallest mole:	$\frac{0.12}{0.12}$ = 1	$\frac{0.12}{0.12}$ = 1

Therefore the empirical formula is CH.

(b). The molecular formula is calculated as follows:

$(CH)n = 78$ (where n is a whole number)

$(12 + 1)n = 78$ (since C = 12 and H = 1)

$\quad\quad 13n = 78$

$\quad\quad\quad n = \dfrac{78}{13}$

$\quad\quad\quad n = 6$

Hence, $(CH)n = (CH)_6 = C_6H_6$.

Therefore the molecular formula is C_6H_6.

9. A hydrocarbon contains 83.3% of carbon. Its density at s.t.p is 6.43g/dm³. Find the:

(a). empirical formula of the hydrocarbon

(b). molecular formula of the hydrocarbon

Solution

(a). The percentage of hydrogen in the hydrocarbon is:

$\quad$ 100 - 83.3 = 16.7%

The empirical formula is obtained as follows:

	C	H
% composition:	83.3	16.7
Ratio of atoms:	$\dfrac{83.3}{12} = 6.94$	$\dfrac{16.7}{1} = 16.7$
Divide by the smallest ratio:	$\dfrac{6.94}{6.94} = 1$	$\dfrac{16.7}{6.94} = 2.4$

This ratio 1 : 2.4 for the atoms, is not a whole number ratio. So we have to make it whole number. In order to make this ratio to be a whole number ratio, we look for the smallest number that can multiply 2.4 to give a whole number. The number is 5. So, multiply 1 and 2.4 by 5 to give:

$\quad$ (1 x 5) : (2.4 x 5)

$\quad$ = 5 : 12

Therefore the empirical formula is C_5H_{12}.

(b). A density of 6.44g/dm^3 means that 1dm^3 of the compound has a mass of 6.43g.

Therefore, at s.t.p: 22.4dm^3 of the compound will have a mass of:

22.4 x 6.43

= 144g

Hence, the molecular mass of the compound is 144

The molecular formula is calculated as follows:

$(C_5H_{12})n$ = 144   (where n is a whole number)

[(12 x 5) + (1 x 12)]n = 78   (since C = 12 and H = 1)

(60 + 12)n = 144

72n = 144

$n = \dfrac{144}{72}$

n = 2

Hence, $(C_5H_{12})n = (C_5H_{12})_2 = C_{10}H_{24}$.

Therefore the molecular formula is $C_{10}H_{24}$.

10. 1.0g of an organic compound containing carbon, hydrogen and oxygen, on oxidation produced 1.38g of carbon (IV) oxide and 1.12g of water. 0.8g vapour of the compound occupied a volume of 0.56dm^3 at s.t.p. Determine:

(a). the empirical formula of the compound

(b). its molecular formula

Solution

(a). Mass of carbon in 1.38g of carbon (IV) is given by:

$\dfrac{12}{44}$ x 1.38   (Note that C = 12 and CO_2 = 44)

= 0.376g

Similarly, the mass of hydrogen in 1.12g of water is given by:

$\frac{2}{18}$ x 1.12   (Note that H_2 = 2 and H_2O = 18)

= 0.124g

Therefore, the mass of oxygen in the compound is given by:

1 - (0.376 + 0.124)   (From the question sample of compound is 1g)

= 1 - 0.5

= 0.5g

We now calculate the empirical formula as follows:

	C	H	O
Mass of element:	0.376	0.124	0.5
Moles of element:	$\frac{0.376}{12}$ = 0.0313	$\frac{0.124}{1}$ = 0.124	$\frac{0.5}{16}$ = 0.0313
Divide by smallest mole:	$\frac{0.0313}{0.0313}$ = 1	$\frac{0.124}{0.0313}$ = 4	$\frac{0.0313}{0.0313}$ = 1

Therefore, the empirical formula is CH_4O

(b). 0.56dm³ of the compound has a mass of 0.8g

Therefore, 22.4dm³ of the compound will have a mass of:

$\frac{22.4}{0.56}$ x 0.8   (This is by simple proportion)

= 32g

Therefore, the molecular mass of the compound is 32g

The molecular formula is calculated as follows:

$(CH_4O)n$ = 32   (where n is a whole number)

(12 + (1 x 4) + 16)n = 32   (since C = 12, H = 1, and O = 16)

$(12 + 4 + 16)n = 32$

$32n = 32$

$n = \dfrac{32}{32}$

$n = 1$

Hence, $(CH_4O)n = (CH_4O)_1 = CH_4O$

Therefore the molecular formula is CH_4O

EXERCISE

1. 2.34kg of an oxide of copper on reduction yielded 2.08kg of copper. Calculate the empirical formula of the oxide (Cu = 63.5, O = 16)

2. On analysis, 3.72g of an oxide of sulphur was found to contain 1.86g of sulphur. Calculate the:

(a). percentage by mass of each element in the compound

(b). empirical formula of the compound

(S = 32, O = 16)

3. The percentage composition of a compound is given by: Sodium 15.9%, Carbon 4.3%, Oxygen 16.7%, and water of crystallization 63.1%. Calculate the formula of the compound. (Na = 23, C = 12, O = 16, H = 1)

4. 10cm^3 of a hydrocarbon is put in a eudiometer tube and sparked with excess oxygen. After the combustion, the volume of the resulting gaseous mixture was found to be 92cm^3. An introduction of a pellet of sodium hydroxide reduced the volume to 52cm^3. After cooling the tube to room temperature, the volume of gas left was found to be 12cm^3. Determine:

(a). the formula of the hydrocarbon

(b). the volume of oxygen sparked with the hydrocarbon

5. 3.6cm^3 of a hydrocarbon is sparked with 75cm^3 of oxygen. After reaction and cooling of the vessel, the volume of the gaseous mixture was found to be 66cm^3. Sodium hydroxide was added to the

reaction vessel. This reduced the gaseous mixture to 48cm³. Determine the formula of the hydrocarbon.

6. 40cm³ of a hydrocarbon was mixed with 250cm³ of oxygen and exploded. After cooling to room temperature, 190cm³ of gas was left. Addition of potassium hydroxide absorbed 120cm³ of the gas. Determine the formula of the hydrocarbon.

7. 12cm³ of a hydrocarbon is sparked with excess oxygen. After the combustion, the volume of the resulting gaseous mixture was found to be 76cm³. An introduction of a pellet of sodium hydroxide reduced the volume to 52cm³. After cooling the tube to room temperature, the volume of gas left was found to be 4cm³. Determine:

(a). the formula of the hydrocarbon

(b). the volume of oxygen sparked with the hydrocarbon

8. 6.24g of a hydrocarbon produced 21.12g of carbon (IV) oxide and 4.32g of water when burnt in air. Find:

(a). the empirical formula of the hydrocarbon

(b). its molecular formula if its molecular mass is 26

(C = 12, O = 16, H = 1)

9. A hydrocarbon contains 83.3% of carbon. Its density at s.t.p is 3.22g/dm³. Find the:

(a). empirical formula of the hydrocarbon

(b). molecular formula of the hydrocarbon

10. 2g of an organic compound containing carbon, hydrogen and oxygen, on burning, produced 2.76g of carbon (IV) oxide and 2.24g of water. 1.6g vapour of the compound occupied a volume of 1.12dm³ at s.t.p. Determine:

(a). the empirical formula of the compound

(b). its molecular formula

CHAPTER 6
EQUILIBRIUM REACTION CALCULATIONS

A reaction is said to be in chemical equilibrium when the rate of the forward reaction is equal to the rate of the backward reaction.

Laws of Chemical Equilibrium

1. If a reversible reaction is represented as $mA + nB \rightleftharpoons pC + qD$, where A and B are the reactants, C and D are the products, while m, n, p, q, are their number of moles respectively, then the equilibrium constant, K_C, is given by:

$$K_C = \frac{[C^p][D^q]}{[A^m][B^n]}$$

K_C is the equilibrium constant in terms of molar concentration, because A, B, C and D in the equation above are concentrations in mol/dm^3 or mol/litre.

2. In the case of gaseous reactions, concentrations are usually expressed in partial pressure. In this case, the equilibrium constant is written as K_P. Let us consider the reaction below.

$$2NO_{(g)} + O_{2(g)} \rightleftharpoons 2NO_{2(g)}$$

The equilibrium constant, K_P, in terms of partial pressure is given by:

$$K_P = \frac{(P_{NO_2})^2}{(P_{NO})^2(P_{O_2})}$$

while the equilibrium constant, K_C, in terms of concentration is given by:

$$K_C = \frac{[NO_2]^2}{[NO]^2[O_2]}$$

For most reactions the values of K_C and K_P are different.

3. Equilibrium constant K, is also related to Gibbs free energy, G. The Gibbs free energy change, ΔG, for a reaction at a particular temperature is given by:

$\Delta G = -RT\ln K$

where R is the molar gas constant, T is the temperature in Kelvin, while K is the equilibrium constant.

Free Gibbs energy change can also be expressed in terms of change in enthalpy, ΔH, and change in entropy, ΔS, as follows:

$$\Delta G = \Delta H - T\Delta S$$

where T is the temperature in Kelvin.

The equilibrium constant, in terms of Gibbs free energy change is given by:

$$K_{eq} = e^{-\Delta G/RT}$$ (This is obtained by making K the subject of the formula $\Delta G = -RT\ln K$)

4. Consider the reaction: $2SO_2 + O_2 \rightleftharpoons 2SO_3$.

Its equilibrium constant is given by:

$$K_1 = \frac{[SO_3]^2}{[SO_2]^2[O_2]}$$

When the reaction above is multiplied by ½, the reaction becomes:

$$SO_2 + \tfrac{1}{2}O_2 \rightleftharpoons SO_3$$

Its new equilibrium constant becomes:

$$K_2 = \frac{[SO_3]}{[SO_2][O_2]^{1/2}}$$

Comparing the expressions for K_1 and K_2 shows that:

$$K_2 = K_1^{1/2}$$

Therefore in general, if a reaction is multiplied by a certain factor, its equilibrium constant must be raised to a power equal to that factor in order to obtain the equilibrium constant of the new reaction.

5. If the equilibrium constant of the forward reaction of a reversible reaction is K_1, then the equilibrium constant K_2, of the backward reaction of the same reversible reaction is given by:

$$K_2 = \frac{1}{K_1}$$

6. Consider the following reactions and their equilibrium constants.

$$2NO + O_2 \rightleftharpoons 2NO_2 \quad K_1 = \frac{[NO_2]^2}{[NO]^2[O_2]}$$

And, $\quad 2NO_2 \rightleftharpoons N_2O_4 \quad K_2 = \dfrac{[N_2O_4]}{[NO_2]^2}$

Adding the two reactions above gives:

$$2NO + O_2 \rightleftharpoons N_2O_4 \quad K_3 = \frac{[N_2O_4]}{[NO]^2[O_2]}$$

Comparing these three equilibrium constants shows that:

$$K_3 = K_2 \times K_1$$

The relationship between K_C and K_P.

The equilibrium constant, K_P, in terms of partial pressure, and the equilibrium constant, K_C, in terms of concentration, are related by:

$$K_P = K_C(RT)^{\Delta ng}$$

where Δng is the number of moles of a gas when going from reactants to products. Δng is given by:

Δng = total number of moles of gaseous products - total number of mole of gaseous reactants

Reaction Quotient, Q_C

The reaction quotient helps us to determine if a reaction has reached equilibrium. It also tells us the direction which a reaction is progressing. The reaction quotient is calculated in the same way as the equilibrium constant.

If the reaction quotient, Q_C = equilibrium constant, K_C, then the reaction is at equilibrium. If $Q_C < K_C$, then the reaction is progressing to the right (forward reaction). If $Q_C > K_C$, then the reaction is progressing to the left (backward reaction).

Examples

1. A reversible reaction is represented as follows:

$$I_2 + I^- \rightleftharpoons I_3^-$$

The concentrations of I_2 and I^- are each equal to 0.02M at the start of reaction. If after reaction the equilibrium concentration of I_2 is 0.008M, what is the equilibrium constant of the reaction?

<u>Solution</u>

We are going to solve this question by establishing an ICE table. ICE means:

    I: **Initial** concentration
    C: **Change** in concentration
    E: **Equilibrium** concentration

In order to create this table, we have to know the change in concentration of each substance in terms of x. From the reaction, the mole ratio of reactants and products is 1 : 1 : 1. Hence the change in concentration are -x, -x, +x. The substance in the direction of the reaction is given positive change as +x, while the substances on the opposite direction are given negative changes, i.e. -x. Therefore the ICE table is given as follows:

ICE	I_2	I^-	I_3^-
Initial concentration:	0.02	0.02	0
Change in concentration:	-x	-x	+x
Equilibrium concentration:	0.02 - x = 0.008	0.02 - x	x

Note that the initial concentration of I_3^- = 0. The equilibrium concentration of each substance is obtained by directly subtracting x from the initial concentration when x is negative (-x), or by directly adding x to the initial concentration when x is positive (+x).

We can solve for x by using the equilibrium concentration of I_2 as follows:

0.02 - x = 0.008   (Note that 0.008 is the equilibrium concentration of I_2 from the question)

Therefore x = 0.02 - 0.008

x = 0.012

Hence the equilibrium concentration of each of the substance is:

I_2 = 0.008

I^- = 0.008   (0.02 - x = 0.02 - 0.012 = 0.008)

I_3^- = 0.012   (i.e. x as shown on the ICE table above)

These values of equilibrium concentrations are now used to calculate the equilibrium constant as follows:

$$K_C = \frac{[I_3^-]}{[I_2][I^-]}$$

$$= \frac{0.012}{0.008 \times 0.008}$$

K_C = 187.5

Therefore the equilibrium constant is 187.5

2. The reversible reaction below shows the reduction of cobalt (II) oxide to produce the metal.

$$CoO + CO \rightleftharpoons Co + CO_2$$

If 1 mole each of CoO and CO are reacted in a 1 litre vessel, equilibrium is attained when 0.4 mole of each of the reactants remains. Calculate the equilibrium constant for the reaction.

Solution

The initial concentration of each of CoO and CO is given by:

$$\text{Concentration} = \frac{\text{number of moles}}{\text{volume}}$$

$$= \frac{1}{1}$$

$$= 1 \text{mol/Litre}$$

The equilibrium concentration of each of CoO and CO is given by:

$$\frac{0.4}{1} = 0.4 \text{mol/L}$$

From the reaction, the mole ration of reactants to products is, 1 : 1 : 1 : 1. Hence the changes in concentration are, -x, -x, +x, +x.. The positive values are assigned to the direction of the reaction. Therefore the ICE table is represented as follows.

ICE	CoO	CO	Co	CO_2
Initial concentration:	1	1	0	0
Change in concentration:	-x	-x	+x	+x
Equilibrium concentration:	1 - x = 0.4	1 - x = 0.4	x	x

Note that the direction of the reaction is assigned the positive changes in concentration i.e. +x.

From the table, 1 - x = 0.4.   (0.4mol/L is the equilibrium concentration of CoO and CO as given in the question).

Therefore, 1 - x = 0.4

$$x = 1 - 0.4$$

$$x = 0.6$$

Hence by using the ICE table above as a guide, the equilibrium concentration of each of the substance is:

$CoO = 0.4$ (As given in the question)

$CO = 0.4$ (As given in the question)

$Co = 0.6$ (i.e. the value of x)

$CO_2 = 0.6$ (i.e. the value of x)

Therefore, the equilibrium constant, K_c, is given by:

$$K_c = \frac{[Co][CO_2]}{[CoO][CO]}$$

$$= \frac{0.6 \times 0.6}{0.4 \times 0.4}$$

$$= 2.25$$

3. Consider the reaction: $N_2 + 3H_2 \rightleftharpoons 2NH_3$ $K_c = 0.05$.

If the equilibrium concentration of nitrogen is 3.4M, and that of hydrogen is 2.2M, calculate the equilibrium concentration of ammonia.

Solution

The equilibrium constant is given by:

$$K_c = \frac{[NH_3]^2}{[N_2][H_2]^3}$$

Substituting known values into the equation above gives:

$$0.05 = \frac{x^2}{3.4 \times (2.2^3)}$$ (where x is the equilibrium concentration of NH_3)

$$0.05 = \frac{x^2}{3.4 \times 10.648}$$

$x^2 = 0.05 \times 3.4 \times 10.648$

$x^2 = 1.81016$

$x = \sqrt{1.81016}$

x = 1.35

Therefore the equilibrium concentration of NH_3 is 1.35M

4. 0.1mole of I_2 and 0.2mole of H_2 are reacted together in a 2 litre vessel to produce HI. If K_C for the reaction is 50, calculate the equilibrium concentration of I_2, H_2 and HI.

Solution

The equation for the reaction is given by:

$H_2 + I_2 \rightleftharpoons 2HI$

The initial concentration of $I_2 = \dfrac{0.2}{2} = 0.05$mol/L

The initial concentration of $H_2 = \dfrac{0.2}{2}$

$= 0.1$mol/L

The initial concentration of HI = 0   (At start of reaction, there is no product)

From the reaction, mole ratio of reactants and product is 1 : 1 : 2. Hence the changes in concentration are -x, -x, +2x. The direction of the reaction is given a positive value. The change in concentration of 2HI is +2x (positive value for direction of reaction). It is 2x since there are 2 moles of HI according to the balanced reaction. These values are now represented using an ICE table as shown below.

ICE	H_2	I_2	HI
Initial concentration:	0.1	0.05	0
Change in concentration:	-x	-x	+x
Equilibrium concentration:	0.1 - x	0.05 - x	x

Therefore, the equilibrium constant is given by:

$$K_C = \dfrac{[HI]^2}{[H_2][I_2]}$$

$$50 = \dfrac{(2x)^2}{(0.1-x)(0.05-x)}$$

$$50 = \frac{4x^2}{0.005 - 0.1x - 0.05x + x^2}$$

$4x^2 = 50(0.005 - 0.1x - 05x + x^2)$

$4x^2 = 0.25 - 5x - 2.5x + 50x^2$

$0 = 50x^2 - 4x^2 - 7.5x + 0.25$

$46x^2 - 7.5x + 0.25 = 0$ (Quadratic equation)

Using quadratic equation formula to solve this equation gives:

$$x = \frac{-(-7.5) \pm \sqrt{(-7.5)^2 - (4 \times 46 \times 0.25)}}{2 \times 46}$$

$$= \frac{7.5 \pm \sqrt{56.25 - 46}}{92}$$

$$= \frac{7.5 \pm \sqrt{10.25}}{92}$$

$$= \frac{7.5 \pm 3.2}{92}$$

$$x = \frac{7.5 + 3.2}{92} \quad \text{Or} \quad x = \frac{7.5 - 3.2}{92}$$

$x = 0.12$ or $x = 0.047$

x cannot be 0.12 since it cannot be greater than any of the initial concentrations.

Therefore x = 0.047

Hence the equilibrium concentrations are

H_2 = 0.1 - x

= 0.1 - 0.047

= 0.053mol/L

Equilibrium concentration of I_2 = 0.05 - x

= 0.003mol/L

Equilibrium concentration of HI = 2x

= 2 x 0.047

= 0.094mol/L

5. Consider the following reaction:

$$2H_{2(g)} + O_{2(g)} \rightleftharpoons 2H_2O_{(l)}$$

If the partial pressure of H_2 and O_2 are 0.95atm and 1.09atm respective, determine the equilibrium constant K_P for the reaction.

<u>Solution</u>

The product $H_2O_{(l)}$, is a liquid and not a gas. Therefore it is not included in the equation for the equilibrium constant. Hence, the equilibrium constant K_P is given by:

$$K_P = \frac{1}{(P_{H_2})^2(P_{O_2})} \quad \text{(The product part not included has been taken to be 1)}$$

$$= \frac{1}{(0.95)^2(1.09)}$$

$$= \frac{1}{0.983725}$$

K_P = 1.02

6. A reaction is represented by: $2N_2O_{5(g)} \rightleftharpoons O_{2(g)} + 4NO_{2(g)}$.

The mole fraction of each substance is given by: N_2O_5: 0.25, O_2: 0.32, and NO_2: 0.43. If equilibrium is established with a total pressure of 2atm, what is the equilibrium constant K_P of the reaction?

<u>Solution</u>

From Raoult's law, the partial pressure of each substance is given by: p = yP, where y is mole fraction, while P is total pressure.

Therefore partial pressure of N_2O_5 = 0.25 x 2   (Total pressure is 2atm)

= 0.5atm

Partial pressure of O_2 = 0.32 x 2

= 0.64atm

Partial pressure of NO_2 = 0.43 x 2

$\qquad$ = 0.86atm

Therefore the equilibrium constant is given by:

$$K_P = \frac{(P_{O_2})(P_{NO_2})^4}{(P_{N_2O_5})^2}$$

$$= \frac{(0.64)(0.86)^4}{0.5^2}$$

$$= \frac{0.64 \times 0.547}{0.25}$$

K_P = 1.40

7. If the total pressure of the reaction system shown below is 6atm, calculate K_P for the reaction.

$\qquad 2Cl_2O_{5(g)} \rightleftharpoons 2Cl_{2(g)} + 5O_{2(g)}$

Solution

Molar mass of $2Cl_2O_5$ = 2[(35.5 x 2) + (5 x 16)]

$\qquad$ = 2(71 + 80)

$\qquad$ = 2 x 151

$\qquad$ = 302g/mol

Molar mass of $2Cl_2$ = 2(35.5 x 2)

$\qquad$ = 2 x 71

$\qquad$ = 142g/mol

Molar mass of $5O_2$ = 5(16 x 2)

$\qquad$ = 5 x 32

$\qquad$ = 160g/mol

Since these molar masses are in g/mol, they can be used to calculate mole fraction as follows:

Total mass in g/mol = 302 + 142 + 160

$$= 604$$

Therefore, mole fraction of $Cl_2O_5 = \dfrac{302}{604}$

$$= 0.5$$

Mole fraction of $Cl_2 = \dfrac{142}{604}$

$$= 0.235$$

Mole fraction of $O_2 = \dfrac{160}{604}$

$$= 0.265$$

We now use each of the mole fractions to calculate partial pressure as follows:

Partial pressure of $Cl_2O_5 = 0.5 \times 6$ (Total pressure is 6atm)

$$= 3 \text{atm}$$

Partial pressure of $Cl_2 = 0.235 \times 6$

$$= 1.41 \text{atm}$$

Partial pressure of $O_2 = 0.265 \times 6$

$$= 1.59 \text{atm}$$

Therefore the equilibrium constant is given by:

$$K_P = \dfrac{(P_{Cl_2})^2 (P_{O_2})^5}{(P_{Cl_2O_5})^2}$$

$$= \dfrac{(1.41)^2 (1.59)^5}{3^2}$$

$$= \dfrac{1.9881 \times 10.162}{9}$$

$K_P = 2.24$

8. Consider the reaction: $½H_{2(g)} + ½I_{2(g)} \rightleftharpoons HI_{(g)}$ $\Delta G° = 1.7 \text{kJ/mol}$ at 25°C.

Calculate the equilibrium constant K for the reaction at 25°C. (R = 8.314J/mol k)

Solution

ΔG° = 1.7 x 1000 = 1700J/mol, T = 25 + 273 = 298k

Therefore the equilibrium constant K is given by:

$$K_{eq} = e^{-\Delta G/RT}$$

$$= e^{-1700/(8.314 \times 298)}$$

$$= e^{-0.686}$$

$$K_{eq} = 0.503$$

9. A reaction is represented as $\frac{3}{2}H_{2(g)} + \frac{1}{2}N_{2(g)} \rightleftharpoons NH_{3(g)}$

ΔH = -45.9KJ/mol, ΔS = -99.2J/mol k at 47°C.

Determine the equilibrium constant of the reaction.

Solution

T = 47 + 273 = 320k

ΔH = -45kJ/mol

$\Delta S = (\frac{-99.2}{1000})$kJ/mol k   (Note that ΔH and ΔS should be expressed in kJ or J)

ΔS = -0.0992kJ/mol k

Recall that: ΔG = ΔH - TΔS

= -45.9 - [320 x (-0.0992)]

= -45.9 + 31.744

ΔG = -14.156kJ/mol

= -14156J/mol   (multiply kJ by 1000 to convert it to J)

Recall that: $K_{eq} = e^{-\Delta G/RT}$

$$= e^{-(-14156)/(8.314 \times 320)}$$

$$= e^{14156/2660.48}$$

$$= e^{5.321}$$

$$K_{eq} = 204.6$$

Therefore the equilibrium constant is 204.6

10. The equilibrium constant for the reaction, $2N_2O_{5(g)} \rightleftharpoons O_{2(g)} + 4NO_{2(g)}$ is 3.2. What is the equilibrium constant for the reaction $N_2O_{5(g)} \rightleftharpoons ½O_{2(g)} + 2NO_{2(g)}$

Solution

$2N_2O_{5(g)} \rightleftharpoons O_{2(g)} + 4NO_{2(g)}$ $K_1 = 3.2$ ....Reaction 1

$N_2O_{5(g)} \rightleftharpoons ½O_{2(g)} + 2NO_{2(g)}$ $K_2 = ?$ ....Reaction 2

Comparing the two reactions above, shows that Reaction 1, has been multiplied by a factor of ½ to obtain reaction 2. Hence the new equilibrium constant will be obtained by raising the old equilibrium constant to a power which is equal to the factor ½.

Therefore $K_2 = K_1^{½}$

$$= 3.2^{½}$$

$$K_2 = 1.79$$

11. Consider the following reaction: $2H_{2(g)} + O_{2(g)} \rightleftharpoons 2H_2O_{(g)}$

If the equilibrium constant for the forward reaction is 2.5, what is the equilibrium constant for the backward reaction?

Solution

Let the equilibrium constant for the forward reaction be K_1, and for the backward reaction be K_2. K_1 and K_2 are related as follows:

$$K_2 = \frac{1}{K_1}$$

$$= \frac{1}{2.5}$$

$$= 0.4$$

Therefore the equilibrium constant of the backward reaction is 0.4.

12. Two reactions and their equilibrium constants are given as follows:

$2NO + O_2 \rightleftharpoons 2NO_2$ $K_1 = 1.85$

$2NO_2 \rightleftharpoons N_2O_4$ $K_2 = 0.92$

From the information provided above, determine the equilibrium constant of the reaction below.

$2NO + O_2 \rightleftharpoons N_2O_4$

Solution

A careful observation of the reactions above shows that the third reaction was obtained by a combination of the first two reactions. Therefore, the equilibrium constant, K_3, of the third reaction can be obtained as follows:

$K_3 = K_1 \times K_2$

$= 1.85 \times 0.92$

$K_3 = 1.702$

13. An equation is represented as follows: $2SO_{2(g)} + O_{2(g)} \rightleftharpoons 2SO_{3(g)}$

If K_C for this equilibrium reaction at 30°C is 8×10^{24}, determine the value of K_P for the reaction at this temperature. (R = 8.314dm³Pa/mol k)

Solution

T = 30 + 273

= 303k

Total gaseous product moles = 2   (From $2SO_3$)

Total gaseous reactants moles = 2 + 1   (From $2SO_2$ and O_2)

$$= 3$$

Therefore, Δn_g = Total gaseous product moles - Total gaseous reactants moles

$$= 2 - 3$$

$$\Delta n_g = -1$$

Hence, $K_P = K_C(RT)^{\Delta n_g}$

$$= 8 \times 10^{24} \times (8.314 \times 303)^{-1}$$

$$= \frac{8 \times 10^{24}}{(8.314 \times 303)^1}$$

$$= \frac{8 \times 10^{24}}{2519.142}$$

$$K_P = 3.18 \times 10^{21}$$

14. Calculate the equilibrium constant, K_C, of the reaction below.

$$CaCO_{3(s)} \rightleftharpoons CaO_{(s)} + CO_{2(g)} \quad K_P = 1.9 \times 10^{-2} \text{ at 500k}$$

(R = 8.314 dm³ Pa/mol k)

Solution

T = 500k

Δn_g = total gaseous product moles - total gaseous reactant moles

$= 1 - 0$ (There is no gaseous reactant, and there is only CO_2 as gaseous product)

$\Delta n_g = 1$

Hence, $K_P = K_C(RT)^{\Delta n_g}$

Therefore, $K_C = \dfrac{K_P}{(RT)^{\Delta n_g}}$

$$= \frac{1.9 \times 10^{-2}}{(8.314 \times 500)^1}$$

$$K_C = 4.57 \times 10^{-6}$$

15. The equilibrium constant for the reaction below is 49.2.

$$H_{2(g)} + I_{2(g)} \rightleftharpoons 2HI_{(g)}$$

If 1.5 moles of H_2 and 1.5 moles of I_2 are placed in a $10 dm^3$ vessel and allowed to react, calculate the concentration of each substance at equilibrium.

Solution

Initial concentration of $H_2 = \dfrac{1.5}{10} = 0.15 mol/dm^3$

Initial concentration of $I_2 = \dfrac{1.5}{10} = 0.15 mol/dm^3$

From the reaction, the mole ratio of reactants to product is 1 : 1 : 2. Hence the changes in concentration are, -x, -x, +2x. We now use these values to present an ICE table as follows.

ICE	H_2	I_2	2HI
Initial concentration:	0.15	0.15	0
Change in concentration:	-x	-x	+2x
Equilibrium concentration:	0.15 - x	0.15 - x	2x

Therefore the equilibrium constant is given by:

$$K_C = \dfrac{[HI]^2}{[H_2][I_2]}$$

Substitute the respective equilibrium concentrations into the equation above. This gives:

$$K_C = \dfrac{(2x)^2}{(0.15 - x)(0.15 - x)}$$

$$49.2 = \dfrac{(2x)^2}{(0.15 - x)^2}$$

Taking the square root of both sides of the equation gives:

$$7.01 = \dfrac{2x}{0.15 - x}$$

$2x = 7.01(0.15 - x)$

$2x = 1.05 - 7.01x$

$2x + 7.01x = 1.05$

$9.01x = 1.05$

$x = \dfrac{1.05}{9.01}$

$x = 0.12$

Therefore the equilibrium concentration of each substance is given by using the ICE table above as a guide as follows.

$H_2 = 0.15 - x$

$= 0.15 - 0.12$

$= 0.03 \text{mol/dm}^3$

$I_2 = 0.03 \text{mol/dm}^3$ (Same as H_2 i.e. $0.15 - x$)

$HI = 2x$

$= 2 \times 0.12$

$= 0.24 \text{mol/dm}^3$

16. A mixture of carbon (II) oxide and steam in the proportion of 2 : 3 by volume is heated to 420°C. If at this temperature the equilibrium constant of the reaction is 1.64, determine the composition of the equilibrium mixture.

Solution

The equation for the reaction is given by: $CO_{(g)} + H_2O_{(g)} \rightleftharpoons CO_{2(g)} + H_{2(g)}$

The volumes given in the question can be used as number of reacting moles since the volume of a gas is proportional to its number of moles. The volume will now serve as our initial concentration. From the reaction, mole ratio is given by, 1 : 1 : 1 : 1. Hence the changes in concentration are, -x, -x, +x, +x. The direction of reaction is given positive values. We now use these values to create an ICE table as follows.

ICE	CO	H_2O	CO_2	H_2
Initial concentration:	2	3	0	0

Change in concentration: −x   −x   +x   +x

Equilibrium concentration: 2 − x   3 − x   x   x

The equilibrium constant is given by:

$$K_C = \frac{[CO_2][H_2]}{[CO][H_2O]}$$

Substituting the respective equilibrium concentrations from the ICE table into the equilibrium constant equation above gives:

$$K_C = \frac{(x)(x)}{(2-x)(3-x)}$$

$$1.64 = \frac{x^2}{6 - 2x - 3x + x^2}$$

$1.64(6 - 5x + x^2) = x^2$

$9.84 - 8.2x + 1.64x^2 - x^2 = 0$

$0.64x^2 - 8.2x + 9.84 = 0$ (Quadratic equation)

$$x = \frac{-(-8.2) \pm \sqrt{(-8.2)^2 - (4 \times 0.64 \times 9.84)}}{2 \times 0.64}$$

$$= \frac{8.2 \pm \sqrt{67.24 - 25.1904}}{1.28}$$

$$= \frac{8.2 \pm \sqrt{42.0496}}{1.28}$$

$$= \frac{8.2 \pm 6.48}{1.28}$$

$$x = \frac{8.2 + 6.48}{1.28} \quad \text{or} \quad x = \frac{8.2 - 6.48}{1.28}$$

x = 11.47   or   x = 1.34

x cannot be 11.47 since it cannot be greater than the initial concentrations of the reactants.

Therefore, x = 1.34

Hence the compositions of the equilibrium mixture in parts by volume are given by:

CO: 2 − x = 2 − 1.34

= 0.66

H_2O: 3 - x = 3 - 1.34

= 1.66

CO_2: x = 1.34

H_2: x = 1.34

17. Consider the following reaction: $CO_{2(g)} + H_{2(g)} \rightleftharpoons CO_{(g)} + H_2O_{(g)}$ K_C = 0.137

At a point during the reaction, the concentration of each substance was found to be: CO_2 = 5M, H_2 = 5M, CO = 1M and H_2O = 1M.

(a). At this point, in which direction is the reaction progressing?

(b). Determine the concentration of each substance at equilibrium

Solution

(a). Let us first calculate the reaction quotient, Q_C. The reaction quotient is calculated in the same way as the equilibrium constant. The reaction quotient helps us to determine if a reaction has reached equilibrium. It also tells us the direction which a reaction is progressing.
If the reaction quotient, Q_C = equilibrium constant, K_C, then the reaction is at equilibrium. If Q_C < K_C (i.e. K_C > Q_C), then the reaction is progressing to the right (forward reaction). If Q_C > K_C (i.e. K_C < Q_C), then the reaction is progressing to the left (backward reaction).

Therefore, $Q_C = \dfrac{[CO][H_2O]}{[CO_2][H_2]}$

$= \dfrac{1 \times 1}{5 \times 5}$

$= \dfrac{1}{25}$

Q_C = 0.04

From the question, K_C = 0.137. This shows that Q_C < K_C, or K_C > Q_C. Therefore, the reaction is progressing to the right. This means that the reaction is in the forward direction.

(b). Let us establish an ICE table for the reaction in order to calculate the equilibrium concentrations of each substance. The ICE table is as shown below.

ICE	CO_2	H_2	CO	H_2O
Initial concentration:	5	5	1	1
Change in concentration:	-x	-x	+x	+x
Equilibrium concentration:	5 - x	5 - x	1 + x	1 + x

The equilibrium constant is given by:

$$K_C = \frac{[CO][H_2O]}{[CO_2][H_2]}$$

$$0.137 = \frac{(1+x)(1+x)}{(5-x)(5-x)}$$

$$0.137 = \frac{1+x+x+x^2}{25-5x-5x+x^2}$$

$$0.137 = \frac{1+2x+x^2}{25-10x+x^2}$$

$1 + 2x + x^2 = 0.137(25 - 10x + x^2)$

$1 + 2x + x^2 = 3.425 - 1.37x + 0.137x^2$

$x^2 - 0.137x^2 + 2x + 1.37x + 1 - 3.425 = 0$

$0.863x^2 + 3.37x - 2.425 = 0$

Using quadratic equation formula, we obtain x as follows:

$$x = \frac{-3.37 \pm \sqrt{(-3.37)^2 - [4 \times 0.863 \times (-2.425)]}}{2 \times 0.863}$$

$$= \frac{-3.37 \pm \sqrt{11.357 + 8.371}}{1.726}$$

$$= \frac{-3.37 \pm \sqrt{19.728}}{1.726}$$

$$= \frac{-3.37 \pm 4.442}{1.726}$$

$$= \frac{-3.37 + 4.442}{1.726}$$

$$= \frac{1.072}{1.726} \quad \text{(The negative value has been neglected since x cannot be negative)}$$

x = 0.621.

Therefore, the equilibrium concentrations are:

CO_2: 5 - x = 5 - 0.621

= 4.379M

H_2: 5 - x = 5 - 0.621

= 4.379M

CO: 1 + x = 1 + 0.621

= 1.621

H_2O: 1 + x = 1 + 0.621

= 1.621

18. Consider the reaction below:

$$NH_4CO_2NH_{2(s)} \rightleftharpoons CO_{2(g)} + 2NH_{3(g)}$$

At room temperature, the total pressure of the gases in equilibrium with the solid is 0.14atm. Calculate the equilibrium constant of the reaction.

<u>Solution</u>

Let us calculate the molar masses of the gaseous products. The solid reactant is not included in the calculation of the equilibrium constant since pressure in involved.

Therefore, $2NH_3$ = 2[14 + (1 x 3)]

= 2(17)

= 34g/mol

CO_2 = 12 + (16 x 2)

= 44g/mol

These masses in g/mol can be used to calculate the mole fraction of each of the gas as follows.

Total mass in g/mol = 34 + 44

$$= 78 \text{g/mol}$$

Therefore, mole fraction of $NH_3 = \dfrac{34}{78}$

$$= 0.436$$

Mole fraction of $CO_2 = \dfrac{44}{78}$

$$= 0.564$$

Therefore by using Raoult's law (p = yP), the partial pressure of each product can be determined as follows:

Partial pressure of $NH_3 = 0.436 \times 0.14$ (Total pressure = 0.14atm)

$$= 0.061 \text{atm}$$

Partial pressure of $CO_2 = 0.564 \times 0.14$

$$= 0.079 \text{atm}$$

Therefore the equilibrium constant is given by:

$K_P = \dfrac{(P_{CO_2})(P_{NH_3})^2}{1}$ (The solid is not included, hence the 1 as our denominator)

$$= \dfrac{(0.079)(0.061)^2}{1}$$

$$= 0.079 \times 0.003721$$

$K_P = 2.94 \times 10^{-4}$

19. Consider the following reaction: $H_{2(g)} + CO_{2(g)} \rightleftharpoons CO_{(g)} + H_2O_{(g)}$ $K_C = 0.2$ at 720k.

If 0.5mole of H_2 and 0.5mole of CO_2 are mixed in a 5 litres vessel at 720k, determine the equilibrium pressure of each substance.

<u>Solution</u>

Initial concentration of $H_2 = \dfrac{0.5}{5}$

$$= 0.1 \text{mol/litre}$$

Initial concentration of CO_2 = $\dfrac{0.5}{5}$

= 0.1 mol/litre

The changes in concentration of reactants and products are, -x, -x, +x and +x, respectively according to the mole ratio of the reaction. Hence the ICE table is given as follows.

ICE	H_2	CO_2	CO	H_2O
Initial concentration:	0.1	0.1	0	0
Change in concentration:	-x	-x	+x	+x
Equilibrium concentration:	0.1 - x	0.1 - x	x	x

Hence the equilibrium constant is given by:

$$K_C = \dfrac{(x)(x)}{(0.1-x)(0.1-x)}$$

$$0.2 = \dfrac{x^2}{(0.1-x)^2}$$

Taking the square root of both sides of the equation gives:

$$0.447 = \dfrac{x}{(0.1-x)}$$

x = 0.447(0.1 - x)

x = 0.0447 - 0.447x

x + 0.447x = 0.0447

1.447x = 0.0447

$$x = \dfrac{0.0447}{1.447}$$

x = 0.0309

Therefore the equilibrium concentration of each substance is given by:

H_2: 0.1 - x = 0.1 - 0.0309

= 0.0691 mol/L

CO_2: 0.1 - 0.0309 = 0.0691mol/L

CO; x = 0.0309

H_2O: x = 0.0309

Recall that: p = mRT, where p = pressure, m = molar concentration, R = 0.0821Latm/mol k, and T = 720.

Therefore pressure of H_2 is: p = mRT

$$= 0.0691 \times 0.0821 \times 720$$

$$= 4.08 \text{atm}$$

Pressure of CO_2 = 4.08atm (The same as that of H_2)

Pressure of CO is: p = mRT

$$= 0.0309 \times 0.0821 \times 720$$

$$= 1.83 \text{atm}$$

Pressure of H_2O: 1.83atm   (Same as CO)

20. A 1 litre flask contains the mixture below in equilibrium.

$$CO + Cl_2 \rightleftharpoons COCl_2$$

CO = 0.05mole, Cl_2 = 0.2mole, $COCl_2$ = 0.2mole.

(a). Calculate the equilibrium constant of the reaction

(b). If 0.1mole of CO is added, what will be the new concentrations of each component when equilibrium is re-established.

Solution

(a) The equilibrium concentration of each substance is given as follows:

CO: $\frac{0.05}{1}$ = 0.05mol/L

Cl_2: $\frac{0.2}{1}$ = 0.2mol/L

$COCl_2$: $\frac{0.2}{1}$ = 0.2mol/L

Therefore the equilibrium constant is calculated from these values as given below.

$$K_c = \frac{[COCl_2]}{[CO][Cl_2]}$$

$$= \frac{0.2}{0.05 \times 0.2}$$

$$= \frac{0.2}{0.01}$$

K_c = 20

(b). When 0.1mole of CO is added to the system, according to Le Chatelier's principle, the system will shift so as to annul this effect. This means that the equilibrium constant will not change. In this case, the previous equilibrium concentration of each substance becomes the initial concentrations. However, the initial concentration of CO will increase to become: 0.1 + 0.05 = 0.15mol/L. The reaction will take place in the forward direction since it was a reactant that was added. Hence from the mole ratio of the reaction components, the changes in concentration will be, -x, -x and +x respectively. We can now present the ICE table as follows:

ICE	CO	Cl_2	$COCl_2$
Initial concentration:	0.15	0.2	0.2
Change in concentration:	-x	-x	+x
Equilibrium concentration:	0.15 - x	0.2 - x	0.2 + x

Hence the equilibrium constant is given by:

$$K_c = \frac{(0.2 + x)}{(0.15 - x)(0.2 - x)}$$

$$20 = \frac{(0.2 + x)}{0.03 - 0.15x - 0.2x + x^2}$$

20(0.03 - 0.35x + x^2) = 0.2 + x

0.6 - 7x + 20x^2 = 0.2 + x

20x^2 - 7x - x + 0.6 - 0.2 = 0

20x^2 - 8x + 0.4 = 0   (Quadratic equation)

Solving this quadratic equation gives:

$$x = \frac{-(-8) \pm \sqrt{(-8)^2 - (4 \times 20 \times 0.4)}}{2 \times 20}$$

$$= \frac{8 \pm \sqrt{64 - 32}}{40}$$

$$= \frac{8 \pm \sqrt{32}}{40}$$

$$= \frac{8 \pm 5.66}{40}$$

$$x = \frac{8 + 5.66}{40} \quad \text{or} \quad x = \frac{8 - 5.66}{40}$$

x = 0.342 or x = 0.0585

But, x cannot be greater than any of the initial concentrations. Therefore, 0.342 cannot be the answer.

Hence, x = 0.0585

Therefore the new concentrations of each component when equilibrium is established again are as given below:

CO: 0.15 - x = 0.15 - 0.0585

= 0.0915mol/L

Cl_2: 0.2 - x = 0.2 - 0.0585

= 0.142mol/L

$COCl_2$: 0.2 + x = 0.2 + 0.0585

= 0.259mol/L

EXERCISE

1. A reversible reaction is represented as follows:

$I_2 + I^- \rightleftharpoons I_3^-$

The concentrations of I_2 and I^- are each equal to 0.1M at the start of reaction. If after reaction the equilibrium concentration of I_2 is 0.02M, what is the equilibrium constant of the reaction?

2. The reversible reaction below shows the reduction of copper (II) oxide to produce the metal.

$CuO + CO \rightleftharpoons Cu + CO_2$

If 2 mole each of CuO and CO are reacted in a 10 litre vessel, equilibrium is attained when 0.9 mole of each of the reactants remains. Calculate the equilibrium constant for the reaction.

3. Consider the reaction: $N_2 + 3H_2 \rightleftharpoons 2NH_3$ $K_C = 0.04$.

If the equilibrium concentration of nitrogen is 1.6M, and that of hydrogen is 0.7M, calculate the equilibrium concentration of ammonia.

4. 1.2mole of I_2 and 1.8mole of H_2 are reacted together in a 5 litre vessel to produce HI. If K_C for the reaction is 52, calculate the equilibrium concentration of I_2, H_2 and HI.

5. Consider the following reaction:

$2H_{2(g)} + O_{2(g)} \rightleftharpoons 2H_2O_{(l)}$

If the partial pressure of H_2 and O_2 are 0.5atm and 1.1atm respective, determine the equilibrium constant K_P for the reaction.

6. A reaction is represented by: $2N_2O_{5(g)} \rightleftharpoons O_{2(g)} + 4NO_{2(g)}$.

The mole fraction of each substance is given by: N_2O_5: 0.45, O_2: 0.12, and NO_2: 0.43. If equilibrium is established with a total pressure of 1.6atm, what is the equilibrium constant K_P of the reaction?

7. If the total pressure of the reaction system shown below is 5atm, calculate K_P for the reaction.

$2Cl_2O_{5(g)} \rightleftharpoons 2Cl_{2(g)} + 5O_{2(g)}$

8. Consider the reaction: $½H_{2(g)} + ½I_{2(g)} \rightleftharpoons HI_{(g)}$ $\Delta G° = 2.1kJ/mol$ at 60°C.

Calculate the equilibrium constant K for the reaction at 60°C. (R = 8.314J/mol k)

9. A reaction is represented as $3/2H_{2(g)} + ½N_{2(g)} \rightleftharpoons NH_{3(g)}$

$\Delta H = -44.3KJ/mol$, $\Delta S = -102J/mol\ k$ at 85°C.

Determine the equilibrium constant of the reaction.

10. The equilibrium constant for the reaction, $2N_2O_{5(g)} \rightleftharpoons O_{2(g)} + 4NO_{2(g)}$ is 0.96. What is the equilibrium constant for the reaction $N_2O_{5(g)} \rightleftharpoons ½O_{2(g)} + 2NO_{2(g)}$

11. Consider the following reaction: $2H_{2(g)} + O_{2(g)} \rightleftharpoons 2H_2O_{(g)}$

If the equilibrium constant for the forward reaction is 1.42, what is the equilibrium constant for the backward reaction?

12. Two reactions and their equilibrium constants are given as follows:

$2NO + O_2 \rightleftharpoons 2NO_2 \quad K_1 = 1.27$

$2NO_2 \rightleftharpoons N_2O_4 \quad K_2 = 1.65$

From the information provided above, determine the equilibrium constant of the reaction below.

$2NO + O_2 \rightleftharpoons N_2O_4$

13. An equation is represented as follows: $2SO_{2(g)} + O_{2(g)} \rightleftharpoons 2SO_{3(g)}$

If K_c for this equilibrium reaction at 25°C is 5×10^{22}, determine the value of K_P for the reaction at this temperature. (R = 8.314dm³Pa/mol k)

14. Calculate the equilibrium constant, K_C, of the reaction below.

$CaCO_{3(s)} \rightleftharpoons CaO_{(s)} + CO_{2(g)} \quad K_P = 2 \times 10^{-4}$ at 410k

(R = 8.314dm³ Pa/mol k)

15. The equilibrium constant for the reaction below is 50.4.

$H_{2(g)} + I_{2(g)} \rightleftharpoons 2HI_{(g)}$

If 2.2moles of H_2 and 2.2moles of I_2 are placed in a 2dm³ vessel and allowed to react, calculate the concentration of each substance at equilibrium.

16. A mixture of carbon (II) oxide and steam in the proportion of 1 : 3 by volume is heated to 600°C. If at this temperature the equilibrium constant of the reaction is 3.8, determine the composition of the equilibrium mixture.

17. Consider the following reaction: $CO_{2(g)} + H_{2(g)} \rightleftharpoons CO_{(g)} + H_2O_{(g)} \quad K_C = 0.096$

At a point during the reaction, the concentration of each substance was found to be: CO_2 = 2M, H_2 = 2M, CO = 0.5M and H_2O = 0.5M.

(a). At this point, in which direction is the reaction progressing?

(b). Determine the concentration of each substance at equilibrium

18. Consider the reaction below:

$$NH_4CO_2NH_{2(s)} \rightleftharpoons CO_{2(g)} + 2NH_{3(g)}$$

At room temperature, the total pressure of the gases in equilibrium with the solid is 0.214atm. Calculate the equilibrium constant of the reaction.

19. Consider the following reaction: $H_{2(g)} + CO_{2(g)} \rightleftharpoons CO_{(g)} + H_2O_{(g)}$ K_C = 1.5 at 540k.

If 1.8mole of H_2 and 1.5mole of CO_2 are mixed in a 5 litres vessel at 300k, determine the equilibrium pressure of each substance.

20. A 5 litre flask contains the mixture below in equilibrium.

$$CO + Cl_2 \rightleftharpoons COCl_2$$

CO = 2moles, Cl_2 = 3moles, $COCl_2$ = 4moles.

(a). Calculate the equilibrium cotrynstant of the reaction

(b). If 2.5moles of Cl_2 is added to the flask, what will be the new concentrations of each component when equilibrium is re-established.